Computational Techniques in Neuroscience

The text discusses the techniques of deep learning and machine learning in the field of neuroscience, engineering approaches to study the brain structure and dynamics, convolutional networks for fast, energy-efficient neuromorphic computing, and reinforcement learning in feedback control. It showcases case studies in neural data analysis.

Features:

- Focuses on neuron modeling, development, and direction of neural circuits to explain perception, behavior, and biologically inspired intelligent agents for decision making
- Showcases important aspects such as human behavior prediction using smart technologies and understanding the modeling of nervous systems
- Discusses nature-inspired algorithms such as swarm intelligence, ant colony optimization, and multi-agent systems
- Presents information-theoretic, control-theoretic, and decision-theoretic approaches in neuroscience
- Includes case studies in functional magnetic resonance imaging (fMRI) and neural data analysis

This reference text addresses different applications of computational neurosciences using artificial intelligence, deep learning, and other machine learning techniques to fine-tune the models, thereby solving the real-life problems prominently. It will further discuss important topics such as neural rehabilitation, brain-computer interfacing, neural control, neural system analysis, and neurobiologically inspired self-monitoring systems. It will serve as an ideal reference text for graduate students and academic researchers in the fields of electrical engineering, electronics and communication engineering, computer engineering, information technology, and biomedical engineering.

Computational Methods for Industrial Applications

Series Editor: Bharat Bhushan

In today's world Internet of Things (IoT) platforms and processes in conjunction with the disruptive blockchain technology and path-breaking artificial intelligence (AI) algorithms lay out a sparking and stimulating foundation for sustaining smarter systems. Further computational intelligence (CI) has gained enormous interests from various quarters in order to solve numerous real-world problems and enable intelligent behavior in changing and complex environment. This book series focuses on varied computational methods incorporated within the system with the help of AI, learning methods, analytical reasoning and sense making in big data. Aimed at graduate students, academic researchers and professionals, the proposed series will cover the most efficient and innovative technological solutions for industrial applications and sustainable smart societies in order to alter green power management, effect of carbon emissions, air quality metrics, industrial pollution levels, biodiversity and ecology.

Blockchain for Industry 4.0: Emergence, Challenges, and Opportunities
Anoop V.S, Asharaf S, Justin Goldston and Samson Williams

Intelligent Systems and Machine Learning for Industry: Advancements, Challenges and Practices
P.R Anisha, C. Kishor Kumar Reddy, Nguyen Gia Nhu, Megha Bhushan, Ashok Kumar, Marlia Mohd Hanafiah

Computational Techniques in Neuroscience

Edited by
Kamal Malik
Harsh Sadawarti
Moolchand Sharma
Umesh Gupta
Prayag Tiwari

CRC Press
Taylor & Francis Group
Boca Raton London New York

CRC Press is an imprint of the
Taylor & Francis Group, an **informa** business

First edition published 2024
by CRC Press
6000 Broken Sound Parkway NW, Suite 300, Boca Raton, FL 33487-2742

and by CRC Press
4 Park Square, Milton Park, Abingdon, Oxon, OX14 4RN

CRC Press is an imprint of Taylor & Francis Group, LLC

ISBN: 978-1-032-46128-1 (hbk)
ISBN: 978-1-032-50343-1 (pbk)
ISBN: 978-1-003-39806-6 (ebk)

DOI: 10.1201/9781003398066

Typeset in Sabon
by MPS Limited, Dehradun

Dr. Kamal Malik would like to dedicate this book to her father, Sh. Ashwani Malik, her mother, Smt. Shakuntla Malik, and her brother, Dr. Shiv Malik, for their constant support and motivation; I would also like to give my special thanks to the publisher and my other co-editors for believing in my abilities. Above all, a humble thanks to the Almighty for this accomplishment.

Dr. Harsh Sadawarti would like to dedicate this book to his father, Sh. Jagan Nath Sadawarti, his mother, Smt. Krishna, and his wife, Ritcha, for their constant support and motivation; I would also like to thank the publisher and my other co-editors for having faith in my abilities. Above all, a humble thanks to the Almighty for this accomplishment.

Mr. Moolchand Sharma would like to dedicate this book to his father, Sh. Naresh Kumar Sharma, and his mother, Smt. Rambati Sharma, for their constant support and motivation, and his family members, including his wife, Ms. Pratibha Sharma, and son, Dhairya Sharma. I also thank the publisher and my other co-editors for believing in my abilities.

Dr. Umesh Gupta would like to dedicate this book to his mother, Smt. Prabha Gupta, and his father, Sh. Mahesh Chandra Gupta, for their constant support and motivation, and his family members, including his wife, Ms. Umang Agarwal, and son, Avaya Gupta. I also thank the publisher and my other co-editors for believing in my abilities. Before beginning and after finishing my endeavor, I must appreciate the Almighty God, who provides me with the means to succeed.

Dr. Prayag Tiwari would like to dedicate this book to his father & his mother for their constant support and motivation and his family members. I also thank the publisher and my other co-editors for believing in my abilities.

Contents

About the Book

Computational neuroscience is an interdisciplinary field that studies the evolution of the nervous system and its working principle, and how it thinks! In computational neuroscience, multi-scale models look at how the brain works, from molecules, cells, and networks to cognition and behavior. It is a way to learn how the nervous system grows and works at different structural levels. Several practitioners and researchers must provide an edge toward artificial intelligence and machine learning concepts to deal with complex real-life problems. Biological neuroscience concepts can be included in different computational paradigms to provide a unique and exciting feature.

Research in computational neuroscience and psychology helps the research community to understand how people act with each other. A psychologist generally perceives a human's feelings or attitude, which decides the person's behaviors. According to computational neuroscientists, the behavior of humans is caused by how well a group of neurons spreads information in a certain way to the brain. As per their understanding, the brain is considered a black box where all connections and processes are abstract. Many inputs and environmental variables are to be fed inside and then processed. As a result, the behaviors of humans will be in action mode. Computational neuroscience tries to discover how the brain works, so identifying people's behaviors before acting will be easy.

Machine learning studies how statistical models and algorithms can make computer systems do things without being directed on how to act. Some computational neuroscientists have explored machine-learning approaches for their research work. However, the research's complexity and cost still need to be improved for machine learning. For example, many computational neuroscience researchers have devised dynamic neural network models over human behavior datasets. The use of machine learning is very insightful in studying how the brain processes and stores information, but soft computing approaches must be noticed for a computational neuroscience study. So, this book will give the ensemble learning of different soft computing approaches and machines and a deep understanding of complex human behavior datasets.

This book explains the effective utilization of different basic computer-devised methods, which will be considered to determine the efficient nervous system. We'll look at how computer science can help to understand visionary control, sensory-motor control, learning, and memory-based tasks. In this book, various specific emerging topics will cover neuron-related information and its network process cycle with adaptability and learning capability. We will also show how to implement using MATLAB, Octave, and Python to make practical neuroscience applications. This will help the reader understand neuroscience-related computing techniques' basic concepts and methods. The book is aimed mainly at third- or fourth-year students in college and beginners-level researchers who want to learn more about how the brain processes information. The edited chapters address action and motor control, neuromodulation, reinforcement learning, vision, and language—the core of human cognition. It provides all necessary neuroscience foundations beyond neurons, synapses, and brain structure and function. The book focuses on neuron modeling, development, and direction of neural circuits to explain perception, behavior, and biologically inspired intelligent agents for decision making.

Preface

We are delighted to launch our book entitled "Computational Techniques in Neuroscience" under the book series Computational Methods for Industrial Applications Series, CRC Press, Taylor & Francis Group. Computational neuroscience is an interdisciplinary field of neural computing, engineering, and artificial intelligence. Various neuroscientists, cognitive researchers, psychologists, and data scientists can publish their work that bridges the gap between neuroscientists and data scientists. Innovative technologies such as artificial intelligence, deep learning, machine learning, and Internet of Things influence today's modern world. This book presents the various approaches, techniques, and applications in computational intelligence and neural engineering. It is a valuable source of knowledge for researchers, engineers, practitioners, and graduate and doctoral students working in the same field. It will also be helpful for faculty members of graduate schools and universities. Around 30 full-length chapters have been received. Among these manuscripts, 11 chapters have been included in this volume. All the chapters submitted were peer-reviewed by at least two independent reviewers and provided with a detailed review proforma. The comments from the reviewers were communicated to the authors, who incorporated the suggestions in their revised manuscripts. The recommendations from two reviewers were considered while selecting chapters for inclusion in the volume. The exhaustiveness of the review process is evident, given the large number of articles received addressing a wide range of research areas. The stringent review process ensured that each published chapter met the rigorous academic and scientific standards.

We would also like to thank the authors of the published chapters for adhering to the schedule and incorporating the review comments. We extend my heartfelt acknowledgment to the authors, peer reviewers,

committee members, and production staff, whose diligent work shaped
this volume. We especially want to thank our dedicated peer reviewers
who volunteered for the arduous and tedious step of quality checking and
critiquing the submitted chapters.

Editor(s)
Kamal Malik
Hash Sadawarti
Moolchand Sharma
Umesh Gupta
Prayag Tiwari

Editor(s)

Dr. Kamal Malik is currently working as a professor in CSE in the School of Engineering and Technology at CTU Ludhiana, Punjab, India. She has published scientific research publications in reputed international journals, including *SCI* and Scopus-indexed journals such as *Adhoc and Senior Wireless Networks, 50, Engineering, Technology and Applied Sciences Research, Journals of Advanced Research in Engineering, Research Journal of Applied Sciences of Engineering & Technology* (RJASET – Maxwell Sciences), *SSRN-Electronic Journal, Design Engineering, Indian Journal of Science and Technology* (IJST), *International Journal of Computer Applications* (IJCA), and many more. She has also attended many national and international conferences of repute like Springer, and Elsevier in India. Her major research areas are artificial intelligence, machine learning and deep learning, data analytics, computational neurosciences, and bio-inspired computing. She has more than 13 years of rich academic and research experience. She has guided 3 research scholars and is currently guiding 8 research scholars at CT University, Ludhiana. She has worked in renowned institutes and universities like RIMT Mandigobindgarh, Maharishi Markandeshwar University, Mullana, GNA University, Phagwara. She has also chaired various sessions in Springer and Elsevier. She has also been awarded the preeminent researcher award from Green Thinkerz Society at CII, Chandigarh. She has completed her doctorate of philosophy in computer science from IKGPTU Kapurthala in 2017, her masters and graduation from Kurukshetra University, Kurukshetra, in 2009 and 2006, respectively.

Dr. Harsh Sadawarti is currently working as vice chancellor of CT University and Professor of Computer Science and Engineering in the School of Engineering and Technology at CTU Ludhiana, Punjab India. He has published scientific research publications in reputed international journals including SCI-indexed and Scopus-indexed journals such as *Journal of Intelligent and Fuzzy Systems, Journal of Applied Sciences, International Journal of Scientific and Technology Research (IJSTR), International Journal of Advanced Research in Engineering and*

Technology (IJARET), International Journal of Light and Electron Optics (Optik Elesevier Journal), International Journal of Computer Science and Communication Engineering (IJCSCE), International Journal of Advanced Research in Computer Science and Software Engineering, Maxwell's Sciences, and many more. Apart from it, he has attended many national, international, and IEEE conferences of repute like Springer, Elsevier in India as well as abroad. He has visited more than 20 countries in his academic career for presenting his scientific research. His major areas of research are machine learning, artificial intelligence, deep learning, parallel processing, computational neurosciences, bio-inspired computing, and security in cloud computing. He has more than 28 years of teaching, academic and research experience in various reputed engineering institutions named as RIMT Mandigobindgarh, Baba Banda Singh Bahadur Institute of Engg and Technology. He has also chaired various international conferences of Springer, Elsevier. He has guided 12 Ph.D. scholars and currently guiding 8 research scholars. He has also been awarded as a Punjab Ratan (Punjab State Intellectuals honor) by All India Conference of Intellectuals at India International Centre New Delhi on 26th of December, 2010. He is also an eminent reviewer of many reputed journals like Elsevier, Springer, etc. He has also been awarded as the best Young Vice Chancellor award from IARE i.e., International Academic and Research Excellence Awards. He has done his doctor of philosophy (Ph.D.) in computer science and engineering and M.Tech (CSE) from Thapar University and his B.Tech(CSE) from Nagpur University in 1999.

Mr. Moolchand Sharma is currently an assistant professor in the Department of Computer Science and Engineering at the Maharaja Agrasen Institute of Technology, GGSIPU Delhi. He has published scientific research publications in reputed international journals and conferences, including SCI-indexed and Scopus-indexed journals such as *Expert Systems* (Wiley), *Cognitive Systems Research* (Elsevier),*Physical Communication* (Elsevier), *Journal of Electronic Imaging* (SPIE), *Intelligent Decision Technologies: An International Journal, Cyber-Physical Systems* (Taylor & Francis Group), *International Journal of Image & Graphics* (World Scientific), *International Journal of Innovative Computing and Applications* (Inderscience), and *Innovative Computing and Communication Journal* (Scientific Peer-reviewed Journal). He has authored/ co-authored chapters with international publishers like Elsevier, Wiley, and De Gruyter. He has authored/edited four books with a national/international level publisher (CRC Press, Bhavya publications). His research areas include artificial intelligence, nature-Inspired computing, security in cloud computing, machine learning, and search engine optimization. He is associated with various professional bodies like IEEE, ISTE, IAENG, ICSES, UACEE, Internet Society, and life membership of the Universal Inovators research lab, etc. He possesses teaching experience more than nine years. He is the co-convener of the ICICC, DOSCI, ICDAM & ICCCN Springer Scopus-Indexed conference series and ICCRDA-2020 Scopus-Indexed IOP Material Science & Engineering conference series. He is also the organizer and co-convener of the International Conference on Innovations and Ideas towards Patents (ICIIP) series. He is also the advisory and TPC committee member of the ICCIDS-2022 Elsevier SSRN Conference. He is also the reviewer of many reputed journals like Springer, Elsevier, IEEE, Wiley, Taylor & Francis Group, IJEECS, and World Scientific Journal, and many Springer conferences. He is also served as a session chair in many international springer conferences. He is a doctoral researcher at DCR University of Science & Technology, Haryana. He completed his Post Graduation in 2012 from SRM UNIVERSITY, NCR CAMPUS, GHAZIABAD, and graduated in 2010 from KNGD MODI ENGG. COLLEGE, GBTU.

Dr. Umesh Gupta is currently an assistant professor at the department of Computer Science, SR University, Warangal, Telengana, India. He received a doctor of philosophy (Ph.D.) (machine learning) from the National Institute of Technology, Arunachal Pradesh, India. He has awarded a gold medal for his master of engineering (M.E.) from the National Institute of Technical Teachers Training and Research (NITTTR), Chandigarh, India, and bachelor of technology (B.Tech.) from Dr. APJ, Abdul Kalam Technical University, Lucknow, India. His research interests include SVM, ELM, RVFL, machine learning, and deep learning approaches. He has published over 35 referred journal and conference papers of international repute. His scientific research has been published in reputable international journals and conferences, including SCI-indexed and Scopus-indexed journals like *Applied Soft Computing* (Elsevier) and *Applied Intelligence* (Springer), each of which is a peer-reviewed journal. His publications have more than 158 citations with an h-index of 8 and an i10-index of 8 on Google Scholar as of March 1, 2023. He is a senior member of IEEE (SMIEEE) and an active member of ACM, CSTA, and other scientific societies. He also reviewed papers for many scientific journals and conferences in the United States and abroad. He led sessions at the 6th International Conference (ICICC-2023), 3rd International Conference on Data Analytics and Management (ICDAM 2023), the 3rd International Conference on Computing and Communication Networks (ICCCN 2022), and other international conferences like Springer ETTIS 2022 and 2023. He is currently supervising two Ph.D. students. He is the co-principal investigator (co-PI) of two major research projects. He published three patents in the years 2021–2023. He also published four book chapters with Springer, CRC.

Dr. Prayag Tiwari received his Ph.D. degree from the University of Padova, Italy. He is currently working as a postdoctoral researcher at Aalto University. Previously, he worked as a Marie Curie Researcher at the University of Padova, Italy. He also worked as a research assistant at the NUST "MISiS", Moscow, Russia. He has several publications in top journals and conferences, including *Neural Networks, Information Fusion,*

IPM, IJCV, IEEE TNNLS, IEEE TFS, IEEE TII, IEEE JBHI, IEEE IOTJ, IEEE BIBM, ACM TOIT, CIKM, SIGIR, AAAI, etc. His research interests include machine learning, deep learning, quantum machine learning, information retrieval, healthcare, and IoT. He is also associated with one funded-based project named "Data Literacy for Responsible Decision-Making," short title (STN LITERACY/Marttinen). He is also the reviewer of many reputed journals like Springer, Elsevier, IEEE, Wiley, Taylor & Francis Group, IJEECS, and World Scientific Journal, and many Springer conferences.

Contributors

Amit Kumar Dutta, Amity Institute of Biotechnology, Amity University Jharkhand, Ranchi, India

Anshika Gupta, Department of Biotechnology, D. Ambedkar Institute of Technology for Handicapped, Kanpur, Uttar Pradesh, India

Analp Pathak, Department of Information Technology, KIET Group of Institutions, Delhi-NCR, Ghaziabad, India

H D Arora, Amity University, Noida, INDIA

Hera Fatma, Department of Biotechnology, Dr. Ambedkar Institute of Technology for Handicapped, Kanpur, Uttar Pradesh, India

Harshit Mishra, Department of Biotechnology, Dr. Ambedkar Institute of Technology for Handicapped, Kanpur, Uttar Pradesh, India

Jeetendra Kumar, Atal Bihari Vajpayee University, Bilaspur, Chhattisgarh, India

Jyoti Sharma, Department of Information Technology, KIET Group of Institutions, Delhi-NCR, Ghaziabad, India

Kalpana Katiyar, Department of Biotechnology, D. Ambedkar Institute of Technology for Handicapped, Kanpur, Uttar Pradesh, India

Kailash Nath Tripathi, Department of AIML, ISBM College of Engineering, Pune, India

Kiran Pal, Delhi Institute of Tool Engineering, Delhi, India

Manisha, Dept of Computer Science, Manipal University, Jaipur, India

Manish Bhardwaj, Department of Computer Science and Information Technology, KIET Group of Institutions, Delhi-NCR, Ghaziabad, India

Mihir Narayan Mohanty, Department of Electronics and Communication Engineering, ITER, Sikhsha 'O' Anusandhan (Deemed to be University), Bhubaneswar, India

M.S.P Subathra, Dept. of Robotics Engineering, Karunya Institute of technology and sciences, Coimbatore, Tamil Nadu, India

Mayank Tyagi, Department of Information Technology, KIET Group of Institutions, Delhi-NCR, Ghaziabad, India

Prasanna J, Dept. of Biomedical Engineering, Karunya Institute of Technology and Sciences, Coimbatore, India

Pratistha Mathur, Department of Information Technology, Manipal University Jaipur, India

Rashmi Gupta, Atal Bihari Vajpayee University, Bilaspur, Chhattisgarh, India

Rekha Jain, Department of Computer Applications, Manipal University Jaipur, India

Sandhyalati Behera, Department of Electronics and Communication Engineering, ITER, Sikhsha 'O' Anusandhan (Deemed to be University), Bhubaneswar, India

S Dhawan, Department of Mathematics, University of Ladakh, Ladakh, India

Surbhi Kumari, Amity Institute of Biotechnology, Amity University Jharkhand, Ranchi, India

Sonu Purohit, Atal Bihari Vajpayee University, Bilaspur, Chhattisgarh, India

Simran Singh, Department of Biotechnology, D. Ambedkar Institute of Technology for Handicapped, Kanpur, Uttar Pradesh, India

S. Thomas George, Dept. of Biomedical Engineering, Karunya Institute of Technology and Sciences, Coimbatore, India

Vijay Kumar, Manav Rachna International Institute of Research Studies, Faridabad, Haryana, India

Vinay Kumar Sharma, School of Computing Science and Engineering, Galgotias University, Greater Noida, India

Yogendra Narayan Prajapati, Department of Computer Science and Engineering, Ajay Kumar Engineering, Ghaziabad, India

Chapter 1

Dynamic intuitionistic fuzzy weighting averaging operator

A multi-criteria decision-making technique for the diagnosis of brain tumor

Vijay Kumar[1], *H D Arora*[2], *and Kiran Pal*[3]

[1]Manav Rachna International Institute of Research Studies,
Faridabad Haryana, India

[2]Amity University, Noida, India

[3]Delhi Institute of Tool Engineering, Delhi, India

1.1 INTRODUCTION

Since the inception of computers, many scientific as well decision support tools have been developed that make it easier for decision makers to make decisions under unfavorable circumstances. Many soft computing techniques such as fuzzy theory and its generalization provide hand held support to decision theory and allow the discipline to grow exponentially in every sphere of the universe. Fuzzy set and its generalizations have applications in many domains, such as decision theory, information theory, medical diagnosis, pattern recognition, etc. Moreover, it acts as an interface and contributes to solving real-life problems with uncertain and imprecise information. Real-life situations are always encountered with deterministic and non-deterministic processes. Out of this, a non-deterministic process is challenging and is taken up by statistics, if found stable. The bitter truth is that many real-life problems are not well defined and are vague; therefore, they are beyond the control of precise notations of mathematics. For such vague situations, Zadeh (1965) introduced the concept of fuzzy set theory, over the generalization of classical set theory, which has the inbuilt ability to represent incomplete information.

From the past five decades, fuzzy theory has grown many folds. Immense research growth has been noticed in the theory of fuzzy sets. Consequently, fuzzy set theory and its generalized versions have emerged as a potential area of interdisciplinary research. Among this, intuitionistic fuzzy set (IFS) theory, introduced by Atanassov (1986), is one such generalization, which is characterized by membership as well as non-membership grades respectively. It is known that IFSs describe the fuzzy character more comprehensively in some special situations and have a variety of applications in decision theory. The complexity of the human body system may not be discussed through traditional approaches, and the behavior of such systems has a certain degree of fuzziness. Zadeh (1969) anticipated that fuzzy theory handles the problems of medical diagnosis very well, as it is a computational tool for dealing with imprecision and uncertainty of human reasoning. The primary characteristic of fuzzy theory is its interpretability that accepts the knowledge in verbal/linguistic ways and allows the system to describe by simple human-friendly rules, which is the key factor in medical discipline. In almost any diagnostic process, medical knowledge is usually imprecise and uncertain and is a relationship between the symptom and the disease. Doctors gather information about the patient from past history, clinical examination, pathological test results, and other investigative procedures such as radiological tests. The knowledge gathered from these sources has a certain degree of incompleteness, and certain things are overlooked or wrongly interpreted during investigation, resulting in wrong treatment of the actual disease. Many fuzzy-based models intervene, varying as per important symptoms, symptom patterns of different diseases, the relationship between diseases and the hypothesis of disease stages, preliminary diagnosis, and final diagnosis. These models form the initial basis for the diseases, which might be iterated and configured as per the requirement. Sanchez (1979) proposed the fuzzy-based model, which represents the knowledge base by establishing a fuzzy max-min relation between the symptom and the disease. In this chapter, medical decision making of various life-threatening diseases has been considered that support doctors and potential patients.

1.2 MULTI-CRITERIA DECISION MAKING

Multi-criteria decision making (MCDM) is the branch of decision making for designing computational tools to support the evaluation of performance criteria with an aim to solve real-world problems of varied disciplines. For the past several decades, MCDM techniques have grown multifold and have provided hand held support to decision makers to classify factors into manageable groups and rank them according to their preferred order. Traditional decision-making problems have certain courses of action with an aim to select the best course of action and are structured into three phases, as given in Figure 1.1.

Atanassov (1986) proposed the concept of IFS, which is the generalization of Zadeh's fuzzy set (1965), characterized by the membership and non-membership functions. Later, Atanassov (1999) and Bustince et al. (2007) developed the theory to handle more complex decision-making problems. De et al. (2000) developed new categories of IFS operators, such as concentration, dilation, and normalization. Based on vague sets and IFS, Chen and Tan (1994), Hong and Choi (2000), Szmidt and Kacprzyk (2002) and Atanassov et al. (2005) developed some approximate techniques for handling MCDM problems under the fuzzy environment. Xu and

```
┌─────────────────────────────────────────────┐
│      Identification and Structuring of Problem │
└─────────────────────────────────────────────┘
                      ⇩
┌─────────────────────────────────────────────┐
│        Development of Model and its Use       │
└─────────────────────────────────────────────┘
                      ⇩
┌─────────────────────────────────────────────┐
│          Development of Action Plans          │
└─────────────────────────────────────────────┘
```

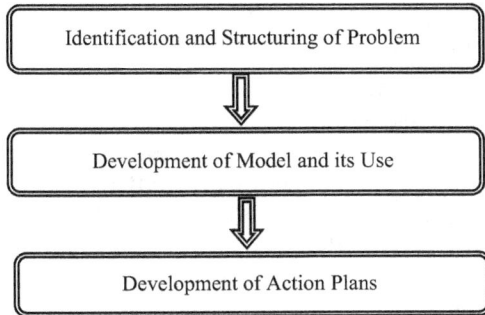

Figure 1.1 Illustrates phases of MCDM problems.

Yager (2006) developed some aggregation operators, which include the intuitionistic fuzzy weighted average (IFWA) operator, intuitionistic fuzzy weighted geometric (IFWG) operator, intuitionistic fuzzy ordered weighted geometric (IFOWG) operator, and intuitionistic fuzzy hybrid geometric (IFHG) operator for handling MCDM problems under the IFS environment. Liu and Wang (2007) developed an evaluation function, commonly known as score functions, to measure the degree to which alternatives satisfy and do not satisfy the requirement. Xu (2007) develop a method for MCDM problems by proposed new IFS preference relations along with their properties. Researchers such as Delgado et al. (1998), Bordogna and Pasi (1993), Fisher (2003), Herrera et al. (1997, 2000a, 2000b, 2000c, 2001), Herrera and Martinez (2000d), Karsak and Tolga (2001), Law (1996), Lee (1996), Roubens (1997), Sanchez (1996), and Yager (1995, 2001) used the application of fuzzy sets in the MCDM problems related to clinical diagnosis of diseases, risk assessment in the development of software, the environment, and manufacturing for decision making. The MCDM problem is a trade-off among the set of factors with respect to the given evaluating conditions. Experts from the same domain involved in the decision making find the best factor among the available ones based on some decision tree under specified criteria with the aim to get the performance rating of the factors obtained by each expert. MCDM models with fuzzy techniques have been proposed by several researchers: Chen and Hwang (1992), Kacprzyk et al. (1992), Fodor and Roubens (1994), Herrera et al. (1996), Bordogna et al. (1997). Hwang and Yoon (1981) pointed out that right assessment of criteria weights is very important in MCDM problems, as the variation of weight values affects the ranking of alternatives.

The information-based computational techniques optimize decisions in each step to bridge the gaps that occur during treatment. These techniques take care of all types of complexities and navigate for patients and doctors by offering other alternatives. Any form of illness present in the human body is deadly, and the trajectory of the disease from diagnosis to treatment passes through certain challenging decisions. Information-based management systems could be the key factor in understanding the diagnosis of diseases, making decisions, and if possible, reconfiguring treatment.

Many decision-making models focus on one criterion and have limited applications; therefore, they fail in most real-life situations. Diaby and Goeree (2014) proposed three stages of the MCDM model, similar to the stages defined in Figure 1.1. Roy (1996) addressed the four types of MCDM problems as: Choice, Sorting, Ranking, and Description. Technique-Ordered Preference by Similarity to the Ideal Solution (TOPSIS) gives promising results and has applications in manufacturing, health care systems, medical decision making, decision making, etc. MCDM strategies tackle complex situations and provide robust solutions to the problem at hand. The TOPSIS method states that the optimal point should be closest to the positive ideal solution and farthest from the negative ideal solution; it yields profit as well as risk as much as possible. TOPSIS is a compensatory method, in which a bad result in one criterion is compensated by a good result in another criterion, with an assumption that each criterion has either a monotonically increasing or decreasing preference function.

1.3 AGGREGATION

The process of representing a collection of data with respect to one representative value is called aggregation. The representative value could be some average maximum or minimum value. In many complex situations, different types of fuzzy and generalized fuzzy-based aggregation operators, such as ordered weighted averaging (OWA) operator, ordered weighted geometric (OWG) operator, weighted harmonic mean (WHM) operator, ordered weighted harmonic mean (OWHM) operator, hybrid harmonic mean (HHM) operator, etc., have been used extensively in different domains of research. For the last couple of years, many MCDM theories and methods have been proposed by using various types of aggregation operators under fuzzy environments with the hypothesis that both the attributes and the decision makers are at the same level of priority. However, in reality, attributes and decision makers have different priority levels. Yager (2008, 2009) proposed a new type of aggregation operator known as prioritized weighted average (PWA) operator. In this chapter, priority operators are used for the diagnosis of the type of brain tumor present in patients, respectively.

1.4 DECISION MAKING

Decision making is a framework by which selection of the best alternative from a finite number of available alternatives has taken place. Moreover, it is an art and a common activity used in every sphere of human functionality and has applications across domains. Decision making receives great interest from researchers across various disciplines, but in real life, the information available is ambiguous or imprecise. To solve such problems of decision making with vague or imprecise information, fuzzy set and IFS theory have emerged as powerful tools. In order to define the entropy function, fuzzy set and IFS theoretic approaches are useful in many real-life situations.

1.5 MEDICAL DIAGNOSIS

Globally, thousands of people die every year due to errors in diagnosis of diseases. Like other domains, the medical domain is characterized by an exponential evolution of knowledge. There are many computational tools related to medical diagnosis, which try to reduce the risk of error and have many advantages. Medical diagnosis begins when a patient consults with a doctor. The doctor evaluates the whole situation of the patient and prepares a knowledge base to prescribe a suitable treatment. The whole process might be iterated and, in each iteration, the diagnosis might be reconfigured, refined, or even rejected. Medical diagnosis is only possible through intensive collaboration between physicians and mathematicians.

1.6 FUZZY THEORY

Fuzzy sets were introduced by Zadeh (1965) as an extension of the classical notion of a set and are defined as:

A fuzzy set A defined in a discrete universe of discourse $X = \{x_1, x_2, \ldots, x_n\}$ is given as: $A = \{< x, \mu_A(x) > : x \in X\}$

Where, $\mu_A: X \to [0, 1]$ is the membership function of set A and $\mu_A(x)$ is called the grade of membership of $x \in X$ in A.

Fuzzy sets have many generalizations like fuzzy soft sets, IFSs, intuitionistic fuzzy soft sets, etc. In this work, IFSs, have been used for the purpose of decision making.

1.7 INTUITIONISTIC FUZZY SETS (IFS)

IFSs were introduced by Atanassov (1986). They are quite useful and applicable, and are defined as:

An IFS A in $X = \{x_1, x_2, \ldots, x_n\}$ is given as: $A = \{< x, \mu_A(x), \nu_A(x) > \mid x \in X\}$ described by membership function $\mu_A(x): X \to [0, 1]$ and non-membership function $\nu_A(x): X \to [0, 1]$ of the element $x \in X$ where the function $\pi_A(x) = 1 - \mu_A(x) - \nu_A(x)$ is defined as an intuitionistic index or hesitation index of x in A. In the limiting case, if $\pi_A(x) = 0$, IFS reduces automatically to a fuzzy set.

1.8 INTUITIONISTIC FUZZY VARIABLE

For a time variable t, $\rho(t) = (\mu_{\rho(t)}, \nu_{\rho(t)})$ is called the intuitionistic fuzzy variable, as proposed by Xu and Yager (2008).

If $t = t_1, t_2, ..., t_p$, then $\rho(t_1), \rho(t_2), ..., \rho(t_n)$ indicate q intuitionistic fuzzy numbers (IFNs) collected at q different periods, where, $\mu_{\rho(t)}, \nu_{\rho(t)} \in [0, 1]$

Some operations of IFNs are as:

Let $\kappa(y_1) = (\mu_{\kappa_1(y_1)}, \nu_{\kappa_1(y_1)}, \pi_{\kappa_1(y_1)})$ and $\kappa(y_2) = (\mu_{\kappa_2(y_2)}, \nu_{\kappa_2(y_2)}, \pi_{\kappa_2(y_2)})$ be two IFNs, then

- $\alpha(y_1) \oplus \alpha(y_2) = (\mu_{\kappa(y_1)} + \mu_{\kappa(y_2)} - \mu_{\kappa(y_1)}\mu_{\kappa(y_2)}, (1 - \mu_{\kappa(y_1)})(1 - \mu_{\kappa(y_2)})$
$- \nu_{\kappa(y_1)}\nu_{\kappa(y_2)})$
- $\lambda\kappa(y_1) = (1 - (1 - \mu_{\kappa(y_1)})^{\lambda}, \nu_{\kappa(y_1)}^{\lambda}, (1 - \mu_{\kappa(y_1)})^{\lambda} - \nu_{\kappa(y_1)}^{\lambda}), \lambda > 0.$

1.9 INTUITIONISTIC FUZZY NUMBER (IFN) AND ITS OPERATIONS

A fuzzy number is a fuzzy set $A \in R$ satisfying the following conditions

    i. A is a convex fuzzy set, i.e., $\lambda x_1 + (1 - \lambda)x_1 \in A$ where x_1, x_2 and $\forall \lambda \in [0, 1]$

    ii. A is a normalized fuzzy set, i.e., $max\ \mu_A(x) = 1$.

    iii. Membership function $\mu_A(x)$ is piecewise continuous over $A \subseteq R$.

    iv. It is defined in a real number.

For $x \in X$, let $A = (\mu_A, \nu_A)$ be the IFN with $0 \leq \mu_A(x) + \nu_A(x) < 1$, where $\mu_A, \nu_A \in [0, 1]$

The operations of IFNs $A_1 = (\mu_{A_1}, \nu_{A_1})$, $A_2 = (\mu_{A_2}, \nu_{A_2})$ and $A_3 = (\mu_{A_3}, \nu_{A3})$ are defined by Yazdani et al. (2016, 2017)

- $A_1 + A_2 = (\mu_{A_1} + \mu_{A_2} - \mu_{A_1}\mu_{A_2}, \nu_{A_1}\nu_{A_2})$
- $A_1 \times A_2 = (\mu_{A_1}\mu_{A_2}, \nu_{A_1} + \nu_{A_2} - \nu_{A_1}\nu_{A_2})$
- $\lambda A = (1 - (1 - \mu_A)^{\lambda}, \nu_A^{\lambda}); \lambda > 0$
- $A^{\lambda} = (\mu_A^{\lambda}, 1 - (1 - \nu_A)^{\lambda}); \lambda > 0$

1.10 DYNAMIC INTUITIONISTIC FUZZY WEIGHTED AVERAGING (DIFWA) OPERATOR

Let $\kappa(y_1), \kappa(y_2), ..., \kappa(y_q)$ be a collection of IFNs collected at q different periods and T be the weight vector of the periods $y_k (k = 1, 2, ..., q)$, then

$DIFWA_{\lambda(y)} \left(\kappa(y_1), \kappa(y_2), \ldots\ldots, \kappa(y_q) \right) = \lambda(y_1)\kappa(y_2) \oplus \ldots\ldots \oplus \lambda(y_q)\kappa(y_q)$ is called dynamic intuitionistic fuzzy weighted averaging (DIFWA) operator.

$$DI_{\lambda(y)} \left(\kappa(y_1), \kappa(y_2), \ldots, \kappa(y_q) \right)$$

$$= \left(1 - \prod_{i=1}^{q} \left(1 - \mu_{\kappa(y_i)} \right)^{\lambda(y_i)}, \prod_{i=1}^{q} v_{\kappa(y_i)}^{\lambda(y_i)}, \prod_{i=1}^{q} \left(1 - \mu_{\kappa(y_i)} \right)^{\lambda(y_i)} - \prod_{i=1}^{q} v_{\kappa(y_i)}^{\lambda(y_i)} \right)$$

where, $\lambda(y_i) \geq 0$; $\sum_{i=1}^{q} \lambda(y_i) = 1$

1.11 MEDICAL DIAGNOSIS OF THE TYPE OF BRAIN TUMOR

As the mechanization of human life activities increases at exponential pace, the risk for the rise of cancer increases. If the treatment of the disease is not started in time, this may be lethal for patients, and the diagnosis of cancer is really a tedious job for doctors. In the era of technology, many evolutionary innovative technologies tackle the problem of diagnosis of diseases and thus increase the efficiency of doctors. Medical diagnosis with the help of computational techniques is a very useful tool for doctors for the purpose of better diagnosis. The idea of MCDM has been extensively used in real-life situations. When it is taken with IFSs, it gives more strength to the concept. Xu and Yager (2006) established certain aggregation operators: IFWG, IFOWG, and IFHG to tackle the problems of multi-attribute decision making. Yager (2004a, 2008, 2009) modeled the concept of prioritized operators to streamline the concept of decision making. Wei (2012) presented the generalized version of prioritized aggregation operators as explained by Yager (2004b, 2009) in a hesitant fuzzy environment and developed some operators of hesitant fuzzy prioritized aggregation. Zhang (2014) presented some prioritized operators under an intuitionistic fuzzy environment, but these operators have certain drawbacks under certain levels. To overcome these limitations, Xu and Yagar (2008) studied the multi-attribute decision-making problems and expressed IFNs at different periods. For the sake of simplicity, such kinds of difficulties as DIF-MADM is governed by the DIFWA operator. In this chapter, an algorithmic approach of the DIFWA operator as proposed by Xu and Yagar (2008) has been discussed and used with ITrFN for the diagnosis of the type of brain tumor over certain attributes. These attributes are given in the form of IFNs. Once the disease has been diagnosed, it is difficult for a doctor to diagnose the type of disease from the available set of diseases. The present work helps the doctors to diagnose the disease and execute the right treatment for the brain tumor and rank it on the basis of available information. For this purpose, we have developed a hypothetical case study to explain the algorithm.

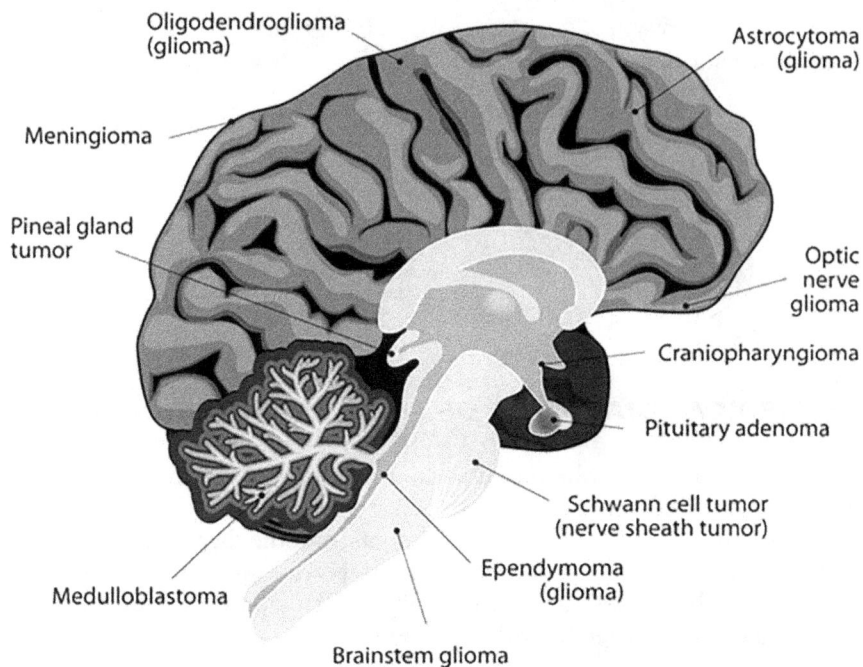

Figure 1.2 Illustrates different types of brain tumors present in human body.

https://www.brainhealthdoctor.com/wp-content/uploads/2018/07/shutterstock_230256940-sm-768×x706.jpg accessed on 20-01-2023.

There are many types of primary brain tumors, named according to the type of the cells or part of the brain in which they grow (Figure 1.2). The most common types are: acoustic neuroma, astrocytoma, brain metastases, choroid plexus carcinoma, craniopharyngioma, embryonal tumors, ependymoma, glioblastoma, glioma, medulloblastoma, meningioma, oligodendroglioma, pediatric brain tumors, pineoblastoma, haemangioblastoma, lymphoma, pineal region tumors, spinal cord tumors, pituitary tumors, germ cell tumors, and many more. Out of these, 78% of malignant tumors belong to gliomas, as they arise from supporting cells of the brain called the glia. Glioma is comprised of glial cells, which protect and support neurons and are of malignant nature, usually located in the brain and spinal cord. Ependymoma is a subtype of glioma generally found in children, which comprises ependymal cells that line the brain's ventricles. According to the National Cancer Institute, approximately 30% of this type of tumor is found in children of age 0–14 years by National Cancer Institute (2018).

Based on factors like: most malignant, aggressive, widely infiltrative, rapid recurrence, and necrosis prone, WHO developed a grading system to judge the malignancy of the tumor as: low grade (Grade I and Grade II) and high grade (Grade III and Grade IV) tumors. Grade I and Grade II tumors generally grow

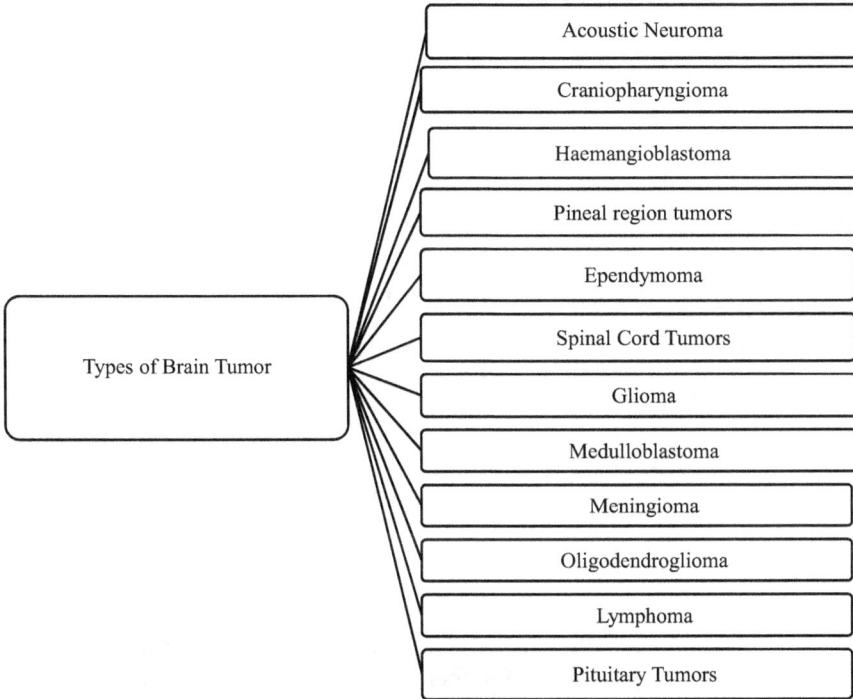

Acoustic Neuroma
Craniopharyngioma
Haemangioblastoma
Pineal region tumors
Ependymoma
Spinal Cord Tumors
Glioma
Medulloblastoma
Meningioma
Oligodendroglioma
Lymphoma
Pituitary Tumors

Types of Brain Tumor

Figure 1.3 Illustrates types of brain tumors.

slower than Grades III and Grade IV tumors and are referred to as low-grade and high-grade tumors. Over a time, a low-grade tumor becomes a high-grade tumor. The grade of a tumor refers to the structure of the cells under a microscope. Brain tumors of any type are treated with surgery, radiation, and/or chemotherapy, either alone or in various combinations. In this chapter, 12 types of brain tumors have been considered and given in Figure 1.3.

1.12 PROPOSED MEDICAL MAKING ALGORITHM: DYNAMIC INTUITIONISTIC FUZZY WEIGHTED AVERAGING (DIFWA) OPERATOR

In this chapter, the DIFWA operator has been used to evaluate the type of brain tumor.

Assumptions:

I. Let F = $\{f_1, \ldots, f_n\}$ be the set of n alternatives.

II. Let S = $\{S_1, \ldots, S_m\}$ be the finite set of attributes, which are collected from different experts and are articulated in IFNs, with $w = (w_1, w_2, \ldots, w_m)^T$ as weight vectors, where $w_j \geq 0$; $\sum_{j=1}^{m} w_j = 1$

III. Let $y_k \ \forall \ (k = 1, 2, ..., q)$ be the q periods whose weight vector is $\kappa(y) = (\kappa(y_1), ..., \kappa(y_q))^T$ as weight vectors; where $\kappa(y_k) \geq 0$; $\Sigma_{k=1}^{q} \kappa(y_k) = 1$

IV. Let $R(y_k) = (t_{ij}(y_k))_{n \times m}$ be an IF decision matrix of the period y_k, where $t_{ij}(y_k) = \left(\mu_{r_{ij}}(y_k), \nu_{r_{ij}}(y_k), \pi_{r_{ij}}(y_k)\right)$ is an attribute value defined by IFN.

Xu and Yagar (2008) proposed the DIFWA operator for solving MCDM problems and is defined as:

12.1 Using the operator(10)

The aggregation of all the IF decision matrix is:

$$T(y_k) = (t_{ij}(y_k))_{m \times n}$$

where, $t_{ij} = (\mu_{ij}, \nu_{ij}, \pi_{ij})$, $\mu_{ij} = 1 - \prod_{k=1}^{k=q}(1 - \mu_{t_{ij}(y_k)})^{\kappa(y_k)}$; $\nu_{ij} = \prod_{k=1}^{k=q} \nu_{t_{ij}(y_k)}^{\kappa(y_k)}$; $\pi_{ij} = 1 - \prod_{k=1}^{k=q}(1 - \mu_{t_{ij}(y_k)})^{\kappa(y_k)} - \prod_{k=1}^{k=q} \nu_{t_{ij}(y_k)}^{\kappa(y_k)}$

12.2. Let $\Omega_i^+ = (\Omega_1^+, ... \Omega_m^+)^T$ and $^-\Omega_i = (\Omega_1^-, ... \Omega_m^-)^T$ be the IF positive ideal solution (IFPIS) and IF negative ideal solution (IFNIS), respectively, where $\Omega_i^+ = (1, 0, 0)$ and $\Omega_i^- = (0, 1, 0)$, $\forall \ (i = 1, 2 .... m)$ be the m largest and m smallest IFNs, respectively. Furthermore, let the alternatives f_i such as $f_i = (t_{i1}, ..., t_{im})^T$, $\forall \ (i = 1, 2.... . n)$.

12.3. Determine the distance between the alternative f_i and the IFPIS(Ω_i^+) and distance between the alternatives f_i and the IFNIS (Ω_i^-), respectively

$$\delta(f_i, \Omega_i^+) = \sum_{j=1}^{m} w_j \delta(t_{ij}, \Omega_j^+) = \sum_{j=1}^{m} w_j(1 - \mu_{ij}) \qquad (1.1)$$

$$\delta(f_i, \Omega_i^-) = \sum_{j=1}^{m} w_j \delta(t_{ij}, \Omega_i^-) = \sum_{j=1}^{m} w_j(1 - \nu_{ij}) \qquad (1.2)$$

12.4. The closeness coefficient of each alternative has been calculated as:

$$C(f_i) = \frac{\delta(f_i, \Omega_i^-)}{\delta(f_i, \Omega_i^+) + \delta(f_i, \Omega_i^-)}, \ \forall \ i = 1, ..., n \qquad (1.3)$$

Since,

$$\delta(f_i, \Omega_i^+) + \delta(f_i, \Omega_i^-) = \sum_{j=1}^{m} w_j(1 + \pi_{ij}) \qquad (1.4)$$

then

$$C(f_i) = \frac{\sum_{j=1}^{m} w_j (1 - \nu_{ij})}{\sum_{j=1}^{m} w_j (1 + \pi_{ij})}, \forall i = 1,...., n \qquad (1.5)$$

12.5. To rank alternative f_i, based on the closeness coefficients $C(f_i)$, the greater the value $C(f_i)$ gives the better alternative.

12.6. End.

1.13 EVALUATION OF CASE STUDY

Let $f_i (i = 1, 2, ..., 11)$ be the available set of various brain tumors described in Figure 1.3. The diagnosis for different brain tumors has been demonstrated on the basis of the prescribed symptoms $C = (c_1, ..., c_5)$ such as different patterns of headaches, unexplained nausea, vision problems, speech difficulties, hearing problems, etc. Also, the decision for the final diagnosis will be made from the panel of decision makers as $D = (d_1, d_2, d_3)$. Let the weight vector for each decision maker be $\lambda(t) = (\frac{1}{6}, \frac{2}{6}, \frac{3}{6})^T$ and the weight vector of the symptoms be $w = (0.1, 0.15, 0.2, 0.25, 0.3, 0.4)^T$. The evaluation of the diagnosis for various brain tumor among $f_i (i = 1, 2, ..., 11)$ is done on the basis of proposed algorithm.

The opinions collected from various experts of the same domain are articulated in IFNs and are listed in the following Tables 1.1–1.3.

Table 1.1 Results of IFN information provided by the expert d_1

Treatment/ Symptom	c_1	c_2	c_3	c_4	c_5
f_1	(0,8,0.1,0.1)	(0.9,0.1,0.0)	(0.7,0.2,0.1)	(0.7,0.2,0.1)	(0.2,0.4,0.4)
f_2	(0.7,0.3,0.0)	(0.6,0.2,0.2)	(0.6,0.3,0.1)	(0.5,0.2,0.3)	(0.2,0.7,0.1)
f_3	(0.5,0.4,0.1)	(0.7,0.3,0.0)	(0.6,0.1,0.3)	(0.4,0.6,0.2)	(0.1,0.8,0.1)
f_4	(0.9,0.1,0.0)	(0.7,0.1,0.2)	(0.8,0.2,0.0)	(0.7,0.1,0.2)	(0.5,0.1,0.4)
f_5	(0.6,0.1,0.3)	(0.8,0.2,0.0)	(0.5,0.1,0.4)	(0.2,0.4,0.4)	(0.4,0.5,0.1)
f_6	(0.3,0.6,0.1)	(0.5,0.4,0.1)	(0.4,0.5,0.1)	(0.2,0.7,0.1)	(0.5,0.5,0.0)
f_7	(0.5,0.2,0.3)	(0.4,0.6,0.0)	(0.5,0.5,0.0)	(0.1,0.8,0.1)	(0.8,0.2,0.0)
f_8	(0,8,0.1,0.1)	(0.9,0.1,0.0)	(0.7,0.2,0.1)	(0.7,0.2,0.1)	(0.5,0.4,0.1)
f_9	(0.7,0.3,0.0)	(0.6,0.2,0.2)	(0.6,0.3,0.1)	(0.5,0.2,0.3)	(0.4,0.6,0.0)
f_{10}	(0.5,0.4,0.1)	(0.7,0.3,0.0)	(0.6,0.1,0.3)	(0.4,0.6,0.0)	(0.6,0.1,0.3)
f_{11}	(0.9,0.1,0.0)	(0.7,0.1,0.2)	(0.8,0.2,0.0)	(0.7,0.1,0.2)	(0.3,0.6,0.1)

Table 1.2 Results of IFN information provided by the expert d_2

Treatment/ Symptom	c_1	c_2	c_3	c_4	c_5
f_1	(0.9,0.1,0.0)	(0.8,0.2,0.0)	(0.8,0.1,0.1)	(0.6,0.3,0.1)	(0.4,0.3,0.3)
f_2	(0.8,0.2,0.0)	(0.5,0.1,0.4)	(0.7,0.2,0.1)	(0.4,0.3,0.3)	(0.7,0.1,0.2)
f_3	(0.5,0.5,0.0)	(0.7,0.2,0.1)	(0.8,0.2,0.0)	(0.7,0.1,0.2)	(0.3,0.5,0.2)
f_4	(0.9,0.1,0.0)	(0.9,0.1,0.0)	(0.7,0.3,0.0)	(0.3,0.5,0.2)	(0.7,0.2,0.1)
f_5	(0.5,0.2,0.3)	(0.6,0.3,0.1)	(0.6,0.2,0.2)	(0.6,0.1,0.3)	(0.8,0.2,0.0)
f_6	(0.4,0.6,0.0)	(0.3,0.4,0.3)	(0.5,0.5,0.0)	(0.2,0.3,0.5)	(0.7,0.3,0.0)
f_7	(0.3,0.5,0.2)	(0.5,0.3,0.2)	(0.6,0.4,0.0)	(0.1,0.5,0.4)	(0.5,0.1,0.4)
f_8	(0.9,0.1,0.0)	(0.8,0.2,0.0)	(0.8,0.1,0.1)	(0.6,0.3,0.1)	(0.7,0.2,0.1)
f_9	(0.8,0.2,0.0)	(0.5,0.1,0.4)	(0.7,0.2,0.1)	(0.4,0.3,0.3)	(0.9,0.1,0.0)
f_{10}	(0.5,0.5,0.0)	(0.7,0.2,0.1)	(0.8,0.2,0.0)	(0.7,0.1,0.2)	(0.5,0.5,0.0)
f_{11}	(0.9,0.1,0.0)	(0.9,0.1,0.0)	(0.7,0.3,0.0)	(0.3,0.5,0.2)	(0.9,0.1,0.0)

Table 1.3 Results of IFN information provided by the expert d_3

Treatment/ Symptom	c_1	c_2	c_3	c_4	c_5
f_1	(0.7,0.1,0.2)	(0.9,0.1,0.0)	(0.9,0.1,0.0)	(0.6,0.1,0.3)	(0.4,0.5,0.1)
f_2	(0.9,0.1,0.0)	(0.6,0.2,0.2)	(0.5,0.2,0.3)	(0.5,0.2,0.3)	(0.7,0.1,0.2)
f_3	(0.4,0.5,0.1)	(0.8,0.1,0.1)	(0.7,0.1,0.2)	(0.3,0.3,0.4)	(0.8,0.2,0.0)
f_4	(0.8,0.1,0.1)	(0.7,0.2,0.1)	(0.9,0.1,0.0)	(0.4,0.4,0.2)	(0.5,0.4,0.1)
f_5	(0.6,0.3,0.1)	(0.8,0.2,0.0)	(0.7,0.2,0.1)	(0.5,0.5,0.0)	(0.9,0.1,0.0)
f_6	(0.2,0.7,0.1)	(0.5,0.1,0.4)	(0.3,0.1,0.6)	(0.1,0.4,0.5)	(0.6,0.1,0.3)
f_7	(0.4,0.6,0.0)	(0.7,0.3,0.0)	(0.5,0.5,0.0)	(0.2,0.3,0.5)	(0.3,0.6,0.1)
f_8	(0.5,0.4,0.1)	(0.7,0.3,0.0)	(0.6,0.1,0.3)	(0.4,0.6,0.2)	(0.1,0.8,0.1)
f_9	(0.9,0.1,0.0)	(0.7,0.1,0.2)	(0.8,0.2,0.0)	(0.7,0.1,0.2)	(0.5,0.1,0.4)
f_{10}	(0.6,0.1,0.3)	(0.8,0.2,0.0)	(0.5,0.1,0.4)	(0.2,0.4,0.4)	(0.4,0.5,0.1)
f_{11}	(0.3,0.6,0.1)	(0.5,0.4,0.1)	(0.4,0.5,0.1)	(0.2,0.7,0.1)	(0.5,0.5,0.0)

The collective results of the experts are listed in Table 1.4.

Using the proposed algorithm, the values of Closeness Coefficients $C(f_i)$ for each type of tumor are given in Table 1.5.

The value of the Closeness Coefficient $C(f_i)$ represents the type of brain tumor.

Table 1.4 Results of collective IFN information provided by the experts $D = (d_1, d_2, d_3)$

Treatment/Symptom	c_1	c_2	c_3	c_4	c_5
f_1	(0.80,0.1,0.094)	(0.87,0.12,0.0)	(0.84,0.11,0.039)	(0.61,0.16,0.219)	(0.37,0.40,0.223)
f_2	(0.84,0.15,0.0)	(0.56,0.15,0.272)	(0.59,0.21,0.192)	(0.46,0.22,0.30)	(0.64,0.13,0.21)
f_3	(0.45,0.48,0.06)	(0.75,0.15,0.09)	(0.72,0.12,0.14)	(0.48,0.23,0.28)	(0.61,0.34,0.04)
f_4	(0.85,0.1,0.04)	(0.79,0.14,0.06)	(0.83,0.16,0.0)	(0.43,0.34,0.22)	(0.57,0.25,0.17)
f_5	(0.56,0.21,0.21)	(0.74,0.22,0.02)	(0.64,0.17,0.18)	(0.49,0.28,0.22)	(0.83,0.16,0.0)
f_6	(0.28,0.64,0.06)	(0.44,0.2,0.35)	(0.39,0.22,0.38)	(0.15,0.39,0.45)	(0.62,0.18,0.18)
f_7	(0.38,0.47,0.14)	(0.60,0.33,0.06)	(0.53,0.46,0.0)	(0.15,0.41,0.43)	(0.49,0.27,0.23)
f_8	(0.74,0.20,0.05)	(0.78,0.21,0.0)	(0.69,0.11,0.19)	(0.53,0.39,0.07)	(0.43,0.44,0.11)
f_9	(0.84,0.15,0.0)	(0.62,0.11,0.26)	(0.74,0.21,0.04)	(0.58,0.16,0.25)	(0.69,0.13,0.16)
f_{10}	(0.55,0.21,0.23)	(0.75,0.21,0.03)	(0.64,0.12,0.22)	(0.45,0.27,0.28)	(0.47,0.38,0.14)
f_{11}	(0.73,0.24,0.02)	(0.73,0.20,0.06)	(0.60,0.36,0.03)	(0.35,0.45,0.19)	(0.69,0.30,0.0)

Table 1.5 Results of closeness coefficients of various types of brain tumors

Type of Tumor	Closeness Coefficients $C(f_i)$
Glioma (f_1)	0.690
Ependymoma (f_2)	0.672
Oligodendrogliom (f_3)	0.654
Meningioma (f_4)	0.692
Pituitary Tumors (f_5)	0.705
Acoustic Neuroma (f_6)	0.539
Craniopharyngioma (f_7)	0.518
Haemangioblastoma (f_8)	0.630
Lymphoma (f_9)	0.727
Pineal Region Tumor (f_{10})	0.622
Spinal Cord Tumors (f_{11})	0.625

1.14   RESULT

This section provides the results with discussion of closeness coefficient for various types of brain tumors, as shown in Figure 1.4.

Thus, the ranking of various brain tumors based on available information is given below:

$$f_9 > f_5 > f_4 > f_1 > f_2 > f_3 > f_8 > f_{11} > f_{10} > f_6 > f_7$$

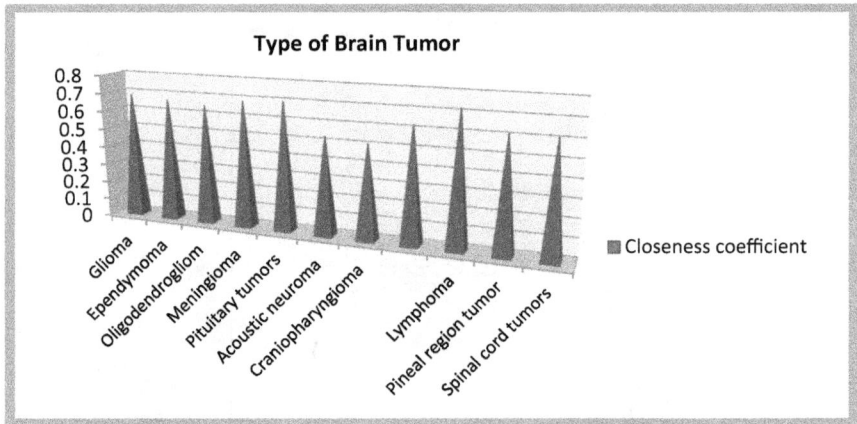

Figure 1.4 Illustrates closeness coefficient for different brain tumors.

1.15 RESULT DISCUSSION

Thus, a better alternative for the above study is f_9 i.e., lymphoma.

This study initially computes the overall collective information of decision makers $D = (d_1, d_2, d_3)$ as given in section 1.5. We find intuitionistic fuzzy ideal solutions, positive as well as negative to calculate distance between IFPIS and IFNIS. To get the optimal decision, closeness coefficient is calculated as given in Table 1.5. The highest value of the closeness coefficient represents the decision value. The result obtained by using algorithm (1.4) is given in Figure 1.4, which gives a clue to the decision makers to make the right decision.

1.16 CONCLUSION

The MCDM technique has been proposed in this chapter for the diagnosis of a type of brain tumor in the context of IFNs. DIFWA operator has been used to aggregate IFNs for each of the alternatives. The optimal decision is evaluated by ranking the alternatives based on their closeness coefficient value. With the help of the given operator, one can easily rank the type of brain tumor present in the body of the patient as per their attributes under the intuitionistic fuzzy environment. This model can be utilized for other methods to get a better decision without doing any additional medical tests, as the proposed algorithm is sufficient enough to perform an initial investigation of the disease. The proposed model may help doctors make better decisions under uncertainty.

REFERENCES

Atanassov, K. "Intuitionistic fuzzy sets". fuzzy sets and systems 20(1), 87–96. DOI: 10.1016/S0165-0114 (86)(1986): 80034-3.

Atanassov, Krassimir T., and Krassimir T. Atanassov. *Intuitionistic fuzzy sets*. Physica-Verlag HD, 1999.

Atanassov, Krassimir, Gabriella Pasi, and Ronald Yager. "Intuitionistic fuzzy interpretations of multi-criteria multi-person and multi-measurement tool decision making." *International Journal of Systems Science* 36, no. 14 (2005): 859–868.

Bordogna, Gloria, and Gabriella Pasi. "A fuzzy linguistic approach generalizing boolean information retrieval: A model and its evaluation." *Journal of the American Society for Information Science* 44, no. 2 (1993): 70–82.

Bordogna, Gloria, Mario Fedrizzi, and Gabriella Pasi. "A linguistic modeling of consensus in group decision making based on OWA operators." *IEEE Transactions on Systems, Man, and Cybernetics-Part A: Systems and Humans* 27, no. 1 (1997): 126–133.

Bustince, Humberto, Francisco Herrera, and Javier Montero, eds. *Fuzzy sets and their extensions: Representation, aggregation and models: Intelligent systems from decision making to data mining, web intelligence and computer vision.* Vol. 220. Springer, 2007.

Chen, Shu-Jen, Ching-Lai Hwang, Shu-Jen Chen, and Ching-Lai Hwang. *Fuzzy multiple attribute decision making methods.* Springer Berlin Heidelberg, 1992.

Chen, Shyi-Ming, and Jiann-Mean Tan. "Handling multicriteria fuzzy decision-making problems based on vague set theory." *Fuzzy sets and systems* 67, no. 2 (1994): 163–172.

De, Supriya Kumar, Ranjit Biswas, and Akhil Ranjan Roy. "Some operations on intuitionistic fuzzy sets." *Fuzzy sets and Systems* 114, no. 3 (2000): 477–484.

Delgado, Miguel, Francisco Herrera, Enrique Herrera-Viedma, and Luis Martinez. "Combining numerical and linguistic information in group decision making." *Information Sciences* 107, no. 1-4 (1998): 177–194.

Diaby, Vakaramoko, and Ron Goeree. "How to use multi-criteria decision analysis methods for reimbursement decision-making in healthcare: a step-by-step guide." *Expert review of pharmacoeconomics & outcomes research* 14, no. 1 (2014): 81–99.

Fisher, Bernard. "Fuzzy environmental decision-making: applications to air pollution." *Atmospheric Environment* 37, no. 14 (2003): 1865–1877.

Fodor, Janos C., and M. R. Roubens. *Fuzzy preference modelling and multicriteria decision support.* Vol. 14. Springer Science & Business Media, 1994.

Herrera, F., E. Herrera-Viedma, and J. L. Verdegay. "A linguistic decision process in group decision making." *Group Decision and Negotiation* 5 (1996): 165–176.

Herrera, F., and E. Herrera-Viedma. "On the linguistic OWA operator and extensions." *The Ordered Weighted Averaging Operators: Theory and Applications* (1997): 60–72.

Herrera, Francisco, and Enrique Herrera-Viedma. "Linguistic decision analysis: steps for solving decision problems under linguistic information." *Fuzzy Sets and Systems* 115, no. 1 (2000a): 67–82.

Herrera, Francisco, and Enrique Herrera-Viedma. "Choice functions and mechanisms for linguistic preference relations." *European Journal of Operational Research* 120, no. 1 (2000b): 144–161.

Herrera, Francisco, Enrique Herrera-Viedma, and Luis Martínez. "A fusion approach for managing multi-granularity linguistic term sets in decision making." *Fuzzy Sets and Systems* 114, no. 1 (2000c): 43–58.

Herrera, Francisco, and Luis Martínez. "A 2-tuple fuzzy linguistic representation model for computing with words." *IEEE Transactions on Fuzzy Systems* 8, no. 6 (2000d): 746–752.

Herrera, Francisco, Enrique Herrera-Viedma, and Francisco Chiclana. "Multiperson decision-making based on multiplicative preference relations." *European Journal of Operational Research* 129, no. 2 (2001): 372–385.

Hong, Dug Hun, and Chang-Hwan Choi. "Multicriteria fuzzy decision-making problems based on vague set theory." *Fuzzy Sets and Systems* 114, no. 1 (2000): 103–113.

Kacprzyk, Janusz, Mario Fedrizzi, and Hannu Nurmi. "Group decision making and consensus under fuzzy preferences and fuzzy majority." *Fuzzy Sets and Systems* 49, no. 1 (1992): 21–31.

Karsak, E. Ertugrul, and Ethem Tolga. "Fuzzy multi-criteria decision-making procedure for evaluating advanced manufacturing system investments." *International Journal of Production Economics* 69, no. 1 (2001): 49–64.

Law, Chiu-Keung. "Using fuzzy numbers in educational grading system." *Fuzzy Sets and Systems* 83, no. 3 (1996): 311–323.

Lee, Huey-Ming. "Applying fuzzy set theory to evaluate the rate of aggregative risk in software development." *Fuzzy sets and Systems* 79, no. 3 (1996): 323–336.

Liu, Hua-Wen, and Guo-Jun Wang. "Multi-criteria decision-making methods based on intuitionistic fuzzy sets." *European Journal of Operational Research* 179, no. 1 (2007): 220–233.

National Cancer Institute. (2018) available at https://www.cancer.gov/types/ childhood-cancers/child-adolescant-cancers-fact-sheet

Roubens, Marc. "Fuzzy sets and decision analysis." *Fuzzy Sets and Systems* 90, no. 2 (1997): 199–206.

Roy, Bernard. *Multicriteria methodology for decision aiding*. Vol. 12. Springer Science & Business Media, 1996.

Sanchez, Elie. "Truth-qualification and fuzzy relations in natural languages, application to medical diagnosis." *Fuzzy Sets and Systems* 84, no. 2 (1996): 155–167.

Sanchez, Elie. "Medical Diagnosis and Composite Fuzzy Relations. Gupta, MM; Ragade, RK; Yager RR (Eds.): Advances in Fuzzy Set Theory and Applications." (1979): 437–444.

Wei, Guiwu. "Hesitant fuzzy prioritized operators and their application to multiple attribute decision making." *Knowledge-Based Systems* 31 (2012): 176–182.

Xu, Zeshui, and Ronald R. Yager. "Some geometric aggregation operators based on intuitionistic fuzzy sets." *International Journal of General Systems* 35, no. 4 (2006): 417–433.

Xu, Zeshui. "Intuitionistic preference relations and their application in group decision making." *Information Sciences* 177, no. 11 (2007): 2363–2379.

Xu, Zeshui, and Ronald R. Yager. "Dynamic intuitionistic fuzzy multi-attribute decision making." *International Journal of Approximate Reasoning* 48, no. 1 (2008): 246–262.

Yager, Ronald R. "An approach to ordinal decision making." *International Journal of Approximate Reasoning* 12, no. 3-4 (1995): 237–261.

Yager, Ronald R. "Fusion of multi-agent preference orderings." *Fuzzy Sets and Systems* 117, no. 1 (2001): 1–12.

Yager, Ronald R. "Modeling prioritized multicriteria decision making." *IEEE Transactions on Systems, Man, and Cybernetics, Part B (Cybernetics)* 34, no. 6 (2004a): 2396–2404.

Yager, Ronald R. "OWA aggregation over a continuous interval argument with applications to decision making." *IEEE Transactions on Systems, Man, and Cybernetics, Part B (Cybernetics)* 34, no. 5 (2004b): 1952–1963.

Yager, Ronald. "Prioritized aggregation operators." *International Journal of Behavioral Science* 16, no. 6 (2008): 538–544.

Yager, Ronald. "Prioritized OWA aggregation." *Fuzzy Optimization & Decision Making* 8, no. 3 (2009): 245–262.

Yazdani, Morteza, Sarfaraz Hashemkhani Zolfani, and Edmundas Kazimieras Zavadskas. "New integration of MCDM methods and QFD in the selection of green suppliers." *Journal of Business Economics and Management* 17, no. 6 (2016): 1097–1113.

Yazdani, Morteza, Prasenjit Chatterjee, Edmundas Kazimieras Zavadskas, and Sarfaraz Hashemkhani Zolfani. "Integrated QFD-MCDM framework for green supplier selection." *Journal of Cleaner Production* 142 (2017): 3728–3740.

Zadeh, Lotfi A. "Fuzzy sets." *Information and Control* 8, no. 3 (1965): 338–353.

Zadeh, Lotfi A. "Biological application of the theory of fuzzy sets and systems." In *The Proceedings of an International Symposium on Biocybernetics of the Central Nervous System*, pp. 199–206. London: Little, Brown and Comp., 1969.

Zhang, Guangquan, and Jie Lu. "Using general fuzzy number to handle uncertainty and imprecision in group decision-making." *Intelligent Sensory Evaluation: Methodologies and Applications* (2004): 51–70.

Chapter 2

Neural modeling and neural computation in a medical approach

Simran Singh, Anshika Gupta, and Kalpana Katiyar
Department of Biotechnology, D. Ambedkar Institute of
Technology for Handicapped, Kanpur, Uttar Pradesh, India

2.1 INTRODUCTION

2.1.1 Introduction

As we all know, biological/spiking neuron models show certain cells in
the nervous system that generate sharp electrical potentials across cell
membranes. But in today's epoch, neuroscientists have developed various
algorithmic and computational approaches to analyze their findings and
make forecasts and hypotheses based on the study's results. However, it is
critical to differentiate between generic modeling and neuronal modeling
related to computational neuroscience. Neural modeling is a mathematical or computer methodology that utilizes a neural network, an artificial
intelligence (AI) technology that trains computers to interpret data in a
manner similar to that of the human brain. Deep learning is a machine
learning approach that engages linked nodes or neurons in a hetero-
structure similar to the human brain. Precise neural models make certain
assumptions according to the available explicit data, and the consequences of these suppositions are quantified. Recent advances in neural
computation reflect multidisciplinary research in theory, statistics in
neuroscience, modeling computation, design, and construction of neurally
inspired information processing systems. Hence, this sector attracts psychologists, neuroscientists, physicists, computer scientists, and AI inves-
tigators functioning on neural systems underlying perception, cognition,
emotion, and behavior and artificial neural systems that have similar
capabilities. Thus, advanced experimental technologies being developed
by brain initiatives will fabricate large, complex data sets and meticulous
statistical analysis and theoretical insight for a better understanding of
these data mean sets.

2.1.2 Why are neuron models better?

A neural or connectionist model is a network of functional processing units
whose local interactions contribute to global model behavior over time.

DOI: 10.1201/9781003398066-2

It is easy to conceive a neural model as having three components: a network, an activation rule, and a learning rule.

Opinions on what defines a decent neuron model differ greatly (Herz et al., 2006). For a long time, two distinct perspectives on this have coexisted: comprehensive biophysical models (proposed in 1952 by neurobiologists Alan Hodgkin and Andrew Huxley) that classify ion channels on the neuronal cell's tree-like spatial configuration, and the "integrate-and-fire" model is a simple mathematical framework used to describe the behavior of neurons in the brain. This model is based on the observation that neurons generate pulsatile electrical activity, also known as action potentials or spikes, through a threshold process. Biophysical models are used by electrophysiologists since they are acquainted with the idea of ion channels that open and close (and hence affect cerebral activity) in response to environmental influences.

On the other hand, theorists often favor uncomplicated neuron models with very few criteria that can be mathematically analyzed. Following prior attempts at the smaller-scale model comparison, the International Neuroinformatics Coordinating Facility (INCF) sponsored an international competition, which enabled a fair measure of neuron models, earlier this year.

The INCF competition is based on the premise that a successful model can predict neuronal activity using data that was not utilized for parameter tweaking. Three in vitro and one in vivo data set were included in the competition. The in vitro data sets were compiled using standard electrophysiological tests in which a random electrical current was delivered into a pyramidal cell and an interneuron through an electrode. Based on their pursuit recorded during the first 38 seconds of data collection, the challenge was to predict the precise moment of a spike in neuronal electrical activity induced throughout a 22-second time frame for 13 (or 9, respectively) repeats of the identical current waveform injected. A basic integrate-and-fire model using a sliding threshold, the winning entry confidently predicted 59.6% (or 81.6%, correspondingly) spike timing of the two neurons.

The majority of threshold models used to study neuronal activity are point neuron models, which assume a simplified neuron construction without any dendritic architecture. However, the INCF competition included a third challenge that involved double-electrode testing to investigate the interaction between somatic and dendritic spike activity. This challenge required the injection of current into both the nerve cell body (soma) and the apical nerve fiber located 600 to 700 micrometers away from the soma. Remarkably, the highest performing model in this challenge was a threshold model that had been augmented with two dendritic equations. The model was able to predict the timing of 90.5% of the spike activity of a neuron in the lateral geniculate area of the brain, given its input triggered by visual stimulation of the retina. This performance was 11%

better than the previous data analysis of the same neuron, highlighting the potential relevance of the competition's results in re-evaluating previous findings. Overall, this challenge demonstrated the importance of considering dendritic architecture in threshold models and the potential for improving our understanding of neuronal activity by incorporating more complex models (Sincich et al., 2009).

Threshold models describe brain function phenomenologically, but they only provide a tenuous relationship to the fundamental biophysical reasons for electrical activity. Threshold models have restricted ability that cannot forecast the specific time course of the voltages before and after a spike, nor can they predict the effect of temperature-dependent, chemical environment changes, or pharmacological modifications of ion channels, whereas Hodgkin-Huxley biophysical models can do all of this. Most biophysical model parameters will soon be assessed in a systematic manner utilizing an appropriate mix of immunostaining methods to estimate ion-channel distribution, calibrated measurements of ion-channel kinetics, and expression investigations to identify tens of ion channels in individual cells. Along those same lines, automatic model construction is feasible.

Furthermore, complex nonlinear spatiotemporal effects on the interaction of back-propagating action potentials (those that move into a dendrite) with shunting inhibition, or local spikes in intracellular calcium concentration that are triggered by numerous, geographically distributed synaptic inputs, are outside the purview of threshold models. Although traditional experimental methods have made it difficult to quantify these nonlinear spatiotemporal aspects, modern imaging techniques that quantify the voltage phase time course from across dendritic tree at a high spatial resolution in tandem with a governed multisite sensory input by glutamate uncaging or optogenetic methods will introduce a fresh era of statistically predictive biophysical models (Gerstner & Naud, 2009).

2.1.3 Objective

We now turn to briefly consider how models of neural networks are being employed in medical computing. Current methods can be divided into two major categories. Neural models are employed as computational tools in one class of applications to execute certain information processing tasks. Neural models are employed as modeling tools in the second category of applications to replicate diverse neurobiological or psychological events.

Aside from its roots in brain modelling and cognitive research, neural models offer a general computational framework with potential applications (Reggia, 1993) in many areas of medical informatics. To the degree that an issue can be described as a neural model, vast computer capacity may be utilized in parallel processing to solve that problem. Furthermore, because neural models have the potential to learn, the knowledge acquisition difficulty experienced when deploying classic AI systems may occasionally be fine-tuned by training an evolving system with realistic test cases. In medical informatics, neural models have been used to solve a range of problems. Image processing is one of these tasks, in which a neural network is utilized to categorize pictures into categories representing therapeutically relevant items. Neural networks, for example, have been taught to identify hepatic ultrasound pictures, as well as to distinguish tumor stages in infrared images and avascular necrosis of the femoral head in magnetic resonance imaging. Another topic that has garnered attention is signal processing. Neural models, for example, have been used to eliminate noise from EKG readings, distinguish evoked potentials, and categorize EMG and electroencephalogram (EEG) waveforms. In the treatment of hypertension, neural models have also been utilized to assist therapeutic decisions such as medication selection and dosage. (Poli et al., 1991) has several further examples of medical uses. In the next section, we will look in depth at one specific application: the use of neural networks for diagnostic issue solving. In a diagnostic problem, one is given specific manifestations (symptoms, signs, and abnormal laboratory test results) and must diagnose the illnesses producing those findings. This type of diagnostic problem solving has been the focus of a great amount of effort in medical informatics over the last 15 years; thus, it provides an excellent field for comparing neural modelling approaches with more classic AI methods.

2.2 DYNAMIC AND ARCHITECTURE FOR NEURAL COMPUTATION

Hopfield in 1982 proposed that the collective features of physical systems may be exploited to directly execute computational tasks, which sparked a lot of interest in brain computation. This new computation paradigm promises to provide new section of computing devices in which the physics of the machine and algorithms of the computation are inextricably linked. The model-independent behavior of dynamical systems will be explored first, and it will be demonstrated that how the systems' phenomenology may be isolated into two fundamental architectural components that conduct the operations of continuous non-linear transformation and auto-associative recall. These filters are primal in the sense that they are essential building pieces from which hierarchical designs can be constructed.

Backpropagation approaches for programming filters will be developed for a basic model, although the concepts are applicable to a wide range of neurodynamical models. The recurrent backpropagation algorithms will be given in a formalism suitable for practical nonlinear dynamical system implementation. One advantage of this formalism is that it employs continuous time and so does not display certain types of oscillations that are linked with backpropagation in models with discrete time (Katiyar et al., 2022).

Among them, one of the model-independent investigation's findings will be used to explain why the backpropagation method cannot store numerous patterns in a basic associative memory model. The problem will be solved by constraining the system during learning. The resultant algorithm not only moves fixed points, but also produces new ones. As a result, autoassociative memory exhibits discontinuous learning behavior.

Two rudimentary filters will be integrated to form an elementary pattern recognition machine as an example of a simple hierarchical design. If the modifications are tiny enough, this machine can recognize patterns that have been corrupted by arbitrary transformations. In other words, the machine can only recognize a limited number of invariant patterns. The two filters in the machine can train separately since error signals do not spread across filter borders. Thus, the two-filter system is a simple illustration of Ballard's modular learning approach (1987)(Pineda, 1988).

2.2.1 Overview of dynamic model

The field of neural network models that can be trained through recurrent backpropagation is very extensive, which is why it's important to focus on a specific system. In this case, we're looking at a neural network algorithm that relies on a set of differential equations.

$$\tau_a \frac{da_i}{dt} = -a_i + \sum_j w_{ij} f(a_i) + K_i \qquad (2.1)$$

These equations involve a state vector called a, an external device vector called K, and an array of weights called w that determine the strength of connections between neurons. The equation also includes a relaxation time scale represented by $-a$, and a differentiable function called $f(a_i)$ that helps determine the system's dynamic properties. Physiologically inspired choices for this function include the logistic and hyperbolic tangent functions. When the weight matrix w is symmetric with zero diagonals, the resulting system is known as a Hopfield model with graded neurons, as described by (Pineda, 1989).

The solutions of equation (2.1) can exhibit oscillations, chaotic behavior, or converge towards isolated fixed points. Since we want to use the fixed point value as the system's output, convergence towards isolated fixed points is the preferred behavior for our purposes. Therefore, when the network is initialized, the weights are adjusted so that the network's output matches the desired output result.

There are several methods available to ensure the convergence of the network, and one such method is to impose a specific structure on the network's connectivity, such as requiring the weight matrix to be lower triangular or symmetric. While symmetry is mathematically elegant, it can be quite strict as it restricts microscopic connectivity by necessitating symmetric coupling between pairs of neurons.

Alternatively, the Jacobian matrix for equation (2.1) can serve as a less stringent criterion for ensuring convergence. The Jacobian matrix can be made bilaterally dominant, which means that the absolute value of the diagonal elements is greater than the sum of the absolute values of the off-diagonal elements in each row. By ensuring the Jacobian matrix is bilaterally dominant, the network can be made stable and converge towards fixed points, even in the absence of strict symmetry constraints.

$$L_{ij} = \delta_{ij} - w_{ij}f'(a_j) \tag{2.2}$$

where δ_{ij} are the elements of the identity matrix and $f'(a_j)$ is the derivative of $f(a_i)$.

Gradient descent dynamics can often result in a neural network that violates stability criteria such as feedforward, symmetric, or diagonal dominance. However, it has been observed that even if the initial network violates these criteria, it does not become unstable during the learning process. This suggests that the stability assumptions underlying recurrent backpropagation are sufficient and that a dynamical system that allows only stable behavior is not necessary.

In gradient descent learning, the objective is to optimize an objective function with the weights as independent parameters. The number of weights, N, is proportional to n^2 if the fan-in/fan-out ratio of the processing units is proportional to R. Relaxing the network and generating a target formula based on the steady-state a^0 takes $O(mN)$ or $O(mn^2)$ procedures. However, computing the gradient of the objective function numerically requires $O(mN^2)$ or $O(mn^{\wedge 4})$ calculations, which becomes impractical for large problems. Moreover, the number of gradient evaluations required for solution convergence may diverge for certain problems.

Backpropagation adaptive dynamics, which is based on gradient descent, uses two methods to reduce computation. The first method represents the gradient of the objective function as an outer-product for equations of the type (2.1), that is,

$$\nabla_w E = b^0 f(a^0)\text{T} \tag{2.3}$$

where a^0 is the solution to equation (2.1) and b^0 is the "error vector,"

$$b^0 = (L^T)^{-1}\text{K} \tag{2.4}$$

The transpose of the n matrix stated in equation (2.2) is denoted by LT, and K is just an external error signal that is dependent on the objective function and a^0. Because L^{-1} can be determined in $O(n^3)$ operations from L and x^2 in

Table 2.1 Results summarization of numerical algorithm complexity

Complexity of Numerical Algorithm	
Worst case (e.g., numerical differentiation)	$O(mN^2)$
Matrix inversion (e.g., gaussian elimination)	$O(mN^{3/2})$
Matrix inversion by relaxation (e.g., recurrent backpropagation)	$O(mN)$
Recursion (e.g., classical feedforward backpropogation	$O(N)$

only $O(mn^2)$ operations, this technique reduces the computational cost of the gradient estimated by a factor of n. As an outcome, the whole computation has scales such as $O(mn^3)$ or $O(mN^{3/2})$. The second approach takes use of the fact that b^0 can be determined via relaxation, or that it is the (stable) fixed point of the linear differential equation.

$$\tau_b \frac{db_i}{dt} = -b_i + f'(a_i) \sum_j w_{ij} b_j + K_i \qquad (2.5)$$

Pineda developed a variant of this equation. Almeida separately developed a discrete time version (1987). It takes $O(n^2)$ operations every time step to relax b (that integrate equation (2.5) until b achieves its steady state). As a result, if the system does not wander chaotically, the amount of computing required scales like $O(mn^2)$ or $O(mn)$. The approach is computationally efficient if the network is big and sparse enough, and the fixed points are not marginally stable. Table 2.1 summarizes these findings. It is worth noting that the two backpropagation algorithms have decreased computation by a factor of N. Because the classical feedforward approach is more efficient since it does not require relaxation to a steady state. (Hong et al., 2007).

2.3 NEURAL MODELING IN FUNCTIONING BRAIN IMAGING

In the last decade, there has been significant advancement in the study of brain function utilizing functioning neuroimaging. It has been notably correct in the realm of human behavior studies in regional cerebral images. The numerous forms of functional neuroimaging approaches are founded on two distinct modalities: (Abeles et al., 1995) *(1) hemodynamic-metabolic* – this domain includes positron emission tomography (PET), single photon emission tomography (SPECT), functional magnetic resonance imaging (fMRI), all predominantly used in humans, and autoradiographic de-oxyglucose and the optical imaging and method, both primarily used in nonhuman animals (Ackermann et al., 1984); and *(2) electric-magnetic* – this realm includes EEG and magnetoencephalography (MEG), both primarily used with human subjects. These two basic forms of imaging have fundamentally distinct properties, the most notable of which are associated with temporal resolution and the amount and type of spatial information provided by each.

2.3.1 Hemodynamic-metabolic methods of functional neuroimaging signal

The study initiated the measurement of brain metabolism and blood flow in humans (Oxide & Of, 1947), whose approach allowed for the measurement of the mean flow of cerebral blood or metabolism for the entire brain. Lassen and Ingvar improved the approach by allowing them to calculate regional cerebral blood flow (rCBF). Several procedures, some less intrusive than others, that were modifications of the Lassen-Ingvar method, were developed throughout the years. (Sokoloff et al., 1977) devised a method for estimating the zonal cerebral metabolic rate for glucose based on the absorption of radioactive deoxyglucose (rCMRglc). Quantitative auto-radiography is used in subhuman animals to quantify radioactive tracer concentrations in the brain. This approach was adapted for use with PET in humans, where radio-labeled fluorodeoxyglucose was used (Horwitz & Sporns, 1994).

2.3.1.1 Functional MRI

Over the last five years, fMRI has become the most widely utilized technology for functional brain imaging (Adey et al., 1961). The most often studied signal is the change in concentration in blood oxygenation and blood volume caused by changing brain activity, which is referred to as BOLD (blood oxygenation level-dependent contrast) (Kwong et al., 1992). Endogenous paramagnetic contrast is provided by deoxygenated hemoglobin. Increase in blood flow lowers the local concentration of deoxygenated hemoglobin, resulting in an increase in the MRI signal on a T2p-weighted picture (Ogawa et al., 1993) These signals, which do not require contrast fluid infusions, can be identified using standard MRI scanners; however, specific gear (e.g., rapid gradient coils) is required. fMRI has a spatiotemporal component.

2.3.1.2 Electric–magnetic methods

Both electric or magnetic fields associated with neuronal activity are measured in the second main kind of functional brain imaging (for reviews, see Gevins et al., 1999). The first functional neuroimaging technologies used on humans captured electrical activity from the scalp. EEGs, which are continual recordings spanning tens to hundreds of seconds, and event-related potentials (ERPs), which are electrical reactions to specific cognitive stimuli; ERPs, which generally correspond to roughly a second's worth of brain activity, are included in this category. Researchers have also been able to track the magnetic fields created by electric current flows associated with cerebral activity paving the way for the application of MEG to study cognitive functions of the brain (Horwitz et al., 2000).

2.3.2 A brief review of neural modeling in functional brain imaging

2.3.2.1 Neuromodeling and PET/fMRI

Neuromodeling has been utilized in three ways in combination with PET and fMRI data (for overviews, see (Horwitz et al., 1999). The first is an investigation into local variations in brain activity, which are translated into alterations in blood flow and metabolism (Motion, 1990).

Secondly, the approach has been used in modelling to identify the systems-level networks that mediate certain cognitive activities (McIntosh & Gonzalez-Lima, 1991) Finally, numerous organizations have begun to build multiple neurobiologically realistic models for simulating PET and fMRI experiments, allowing one to compare system-level results to neuronal and neural ensemble-level results (Goodisman & Asmussen, 1997).

2.3.2.2 EEG/MEG and neuromodeling

The first neural modelling experiments utilizing functional brain imaging data appeared to be concentrated on EEG data. Given that EEG was the initial neuroimaging modality, this is not surprising (Brasil-Neto et al., 1992). The neurophysiological underpinning for the EEG/ERP signals was a major issue. The fundamental mechanics are now rather well known. The extracellular environment is affected by ionic currents, which flow through the neuronal membrane that are sinks as well as sources. When several comparable parts of an anatomically structured ensemble of neurons are engaged at the same time, these sinks and sources (dipoles) can reach macroscopic size. Neuronal modelling of such ensembles led to the conclusion in the cortex, which is the activity of postsynaptic potentials linked with pyramidal neurons that is important, rather than the activity of pyramidal neurons themselves.

2.3.3 Conclusion

The field of functional brain imaging is a significant source of complicated data in neuroscience study. Temporal as well as spatial domains are complicated. Therefore, these data enable researchers to look into the neurological underpinnings of human sensory, motor, affective, and cognitive functions. The secondary argument is that this complexity limits tranquil comprehension and necessitates a similarly rich computational approach to data analysis and, more importantly, data interpretation. The tertiary part is that computational modeling of functional imaging data can occur at different levels (microscopic, mesoscopic, and macroscopic), which also leads to the fourth point that bridging models will be needed to connect all of these approaches into unified and accordant accounts of brain function (Horwitz et al., 2000).

2.4 LITERATURE REVIEW

2.4.1 Type of neural model

There are three types of neural models that are described below:

2.4.1.1 Single cell level models

Dendritic tree modelling at single nerve cells is an essential method that is extensively utilized for examining all the features of nerve cells (Horwitz et al., 1998) Such investigations often highlight the interaction of several biophysical factors that are related to giving out synaptic inputs also accompanied by dendritic shape for the development of a response in the modelled neuron. For example, Traub (1982) explored the development of electro-physiologically visible exploding patterns within CA3 hippocampal pyramidal neurons using compartmental modelling and discovered essential variables. A new model of neurons in the piriform cortex demonstrated in what manner neuro-pharmacological action (acetylcholine release) might alter pattern storage. A comprehensive model about the cortical neuropil that included dendritic shape along with fundamental interrelationship in the middle of neurons marked a variety of issues starting with neural evolution to biochemical trace generation associated with neural action (Gally et al., 1990). Nitric oxide is utilized as a free diffusing extracellular messenger in the model. This type of "tissue modelling" lies down between the ensemble and single cell model.

2.4.1.2 Ensemble-level models

Models comprising tens to thousands of nerve cells linked to create networks address a distinct set of problems. These models may show how inhibitory & excitatory effects are combined in aggregative forms about neuronal action. For vertebrate & invertebrate central motif generators, detailed and realistic models have been created.

Freeman's investigations about the olfactory bulb of rats in higher vertebrate brains have provided crucial awareness about electroencephalographic signals collected above the bulb's plane are formed by the bulb's constituent neurons (Jindra, 1976). This and comparable olfactory bulb computational models have been highly effective in describing a wide range of actual data. They extended their effort at the basis of single cells by creating anatomically and physiologically accurate model of groupings of hippocampal neurons. They are able to demonstrate how harmonic action is created among the hippocampus, as well as why their anatomical & physiological characteristics at the level of neurons are important. Their computer investigations were closely related to actual experimental data. Neocortical network models have demonstrated in what way a bunch of neurons function broadly & how the action of a collective single neuron converts within a macroscopically visible quantity like local field capacity (Seemanthini & Manjunath, 2021).

2.4.1.3 Systems-level models

Neurological model of entire brain areas or complete nervous systems try to reveal the nerve foundation for effective processes like behavioral or cognitive tasks. Cognitive models and neural models differ in their approach to studying the brain and its functions. While neural models focus on describing the underlying neural mechanisms that give rise to behavior, cognitive models attempt to explain behavior in terms of abstract mental processes or representations (Cowan, 1998).

Systems-level neural models are often connected to simulated or actual behaviors, which may be used to objectively analyze such a model's performance in the context of a specific task. There have been few attempts to formally model whole cortical regions or groupings of areas. The relative youth of this sort of study is due, in part, to the astonishing computational needs of a genuine neural model at about this scale, and, in part, to the conceptual difficulties of identifying important structural and dynamic constraints to also be incorporated in such a model. Existing models often employ crude approximations of single neural units, typically representing complete neuronal circuits or populations rather than single cells.

2.4.2 Machine learning

Machine learning enables machines to learn without being explicitly programmed. What does "without being expressly programmed" mean? To understand this, consider the typical programming technique, in which we explicitly program everything and provide it to the machine, and the machine acts or delivers output based on the program. Traditional programming examples include calculator programs, finding factorials, determining if a value is odd or even, and so on. If we give the machine an addition logic, it will constantly execute addition; it will never learn to conduct subtraction until expressly programmed to do so. The machine is not learning in this case; it just produces the output that has been programmed (Srivastava et al., 2021).

We need two things to properly apply machine learning: data and an algorithm. These are the absolutely necessary conditions for implementing machine learning. Appropriate data aids in the generation of more accurate models. What does "suitable" imply in this context? For example, if our goal is to categorize students based on grades, the relevant data would include academic data from students, which will aid in better categorizing students based on grades. However, if we offer students' personal information data instead of academic data, the model developed will not be useful in achieving our goal.

2.4.2.1 Artificial intelligence vs machine learning vs deep learning

2.4.2.1.1 Artificial intelligence

While AI is a field of science that involves creating machines that can perform tasks that would normally require human intelligence, it is not necessarily focused solely on creating robots. In fact, AI can be used to develop a wide range of intelligent systems, including software programs and digital assistants, as well as physical robots.

There are two stages of AI:

a. **General Artificial Intelligence:** This can execute any intellectual work as accurately as a person.
b. **Narrow Artificial Intelligence:** This can outperform humans in a certain activity.

2.4.2.1.2 Machine learning

Machine learning is a subsection of AI in which we attempt to equip machines with the capacity to learn on their own without being explicitly programmed. To forecast/predict the output in machine learning, we apply statistical algorithms including decision tree, support vector machine), and KNN.

There are three types of machine learning categories:

2.4.2.1.2.1 SUPERVISED LEARNING

"Machine learns under supervision," as the name indicates, and "train me" to make it easier for you. Consider a student studying for test questions and having access to the answer key to all those questions; in this scenario, the student is our model, the exam questions are the input, and the solution key is the intended output. This form of data is referred to as "labeled data," and it is used in supervised learning. We train the model using labeled data, and the model learns the link between both the output and input variables/features during training. As the training phase concludes, the model is tested and therefore enters the testing phase, in which the test input characteristics are supplied to the model, and the model will now categorize the output/predicted output. Now that the anticipated and intended outputs have been matched, we can claim that the model is more appropriate and the error margin is low; however, if the gap between the two is more, we can say that the error margin is greater and the model must be trained more/better. So in supervised learning:

- Labeled data should be used to train the model.
- Feedback is provided for the model to improve.
- Predicting/classifying the output.

2.4.2.1.2.2 UNSUPERVISED LEARNING

"Machine learns on its own without supervision," as the name implies. The output variables in unsupervised learning are unlabeled. There are no known combos of input and output variables. Unsupervised learning is concerned with examining correlations between input variables and identifying hidden patterns that may be used to generate new labels for potential outputs. For example, a student is given 50 shapes, but the kids have no clue what they are (they are still learning), and we do not define any labels or names of the forms. This is an example of unsupervised learning with unlabeled data. Now the student will try to understand the patterns; for example, the student will make a group of shapes that has four comers, another group of shapes with three corners, and one more group of shapes with two corners. So, here the student tried to make clusters of similar input elements, and that's what we do in unsupervised learning. Further to those clusters made by students, new labels could be given. You are right if you think that the shape labels/names are quadrilateral, triangle, and circle. So, in unsupervised learning:

- Unlabeled data is provided.
- No feedback is given.
- Finds hidden patterns in data.

2.4.2.1.2.3 REINFORCEMENT LEARNING

In contrast to supervised and unsupervised learning, reinforcement learning constructs a prediction model by gathering input from random "trial and error" as well as learning from previous iterations. We can also say that reinforcement learning is a prize; for example, in the case of self-driving cars, avoiding a crash results in a positive score, while crashing results in a negative score; from this, the model learns which actions should be executed under what conditions in the environment.

2.4.2.1.3 Deep learning

Deep learning is a subdivision of machine learning. It is the next footstep in progression of machine learning. It has a layered architecture and applies an artificial neural network (ANN), which was motivated more by biological neural networks. The human brain often evaluates and processes the information it gets before attempting to recognize it from earlier stored knowledge. Deep learning algorithms are similarly trained to recognize patterns and categorize various forms of information in order to produce the required output when given an input. In machine learning, we must manually give the features. However, deep learning automatically extracts features for categorization, which necessitates a large quantity of data for training the deep learning algorithm. As a result, the accuracy of deep learning output is dependent on the quantity of data. We employed convolutional neural networks (CNN), ANN, and recurrent neural networks (RNN) in deep learning.

2.4.3 Application of machine learning

Machine learning is an interdisciplinary discipline that may be utilized in a variety of fields such as science and business.

Machine learning is applied in medical science to detect and diagnose disorders. As a result, many diseases are treated and medical technology is speedily evolving. We have the ability to create three-dimensional images that can anticipate the exact location of a lesion inside the brain. It makes possible detection of brain cancers as well as other brain-related illnesses, image recognition, traffic prediction, speech recognition, product recommendations, email spam and malware filtering, self-driving cars, virtual personal assistant, stock market trading, online fraud detection, automatic language translation (Srivastava et al., 2021).

Machine learning implementations are not restricted to the examples given here; there are several more areas where machine learning has proven its worth, such as picture identification, speech recognition, medical diagnosis, and learning connections.

It is a phenomenal breakthrough in the area of AI. These machine learning applications are just a few examples of how this technology may enhance our lives.

2.4.3.1 Machine learning in healthcare

Machine learning is critical in health care. The following are some data on this subject:

- Disease Identification
- Drug Discovery
- Personalized Treatment
- Clinical Trial Results
- Smart Electronic Health Records
- Treatment Queries and Suggestions

2.4.4 Types of algorithms being used

2.4.4.1 Logistic regression

Logistic regression is also another approach adopted from statistics by machine learning. It is the preferred strategy for binary classification issues (those with only two class values) (By, 2018). There are two sorts of variables in logistic regression: dependent variables and independent variables. The dependent variables are binary, which means they are either 1 (true, success) or 0 (false, failure). The purpose of this technique is to discover the model that best fits the connection between the independent and dependent variables. Independent variables can be either continuous or binary in nature. It is also known as logistic regression. This can deal with the likelihood of measuring the relationship between the dependent and independent variables.

2.4.2.2 Convolutional neural network

CNN is among the most widely used deep technique based on the animal's visual cortex (Levitt et al., 1994). CNN is now widely utilized in object identification and tracking (Fan et al., 2010), text recognition and detection, visual detection, posture estimation (Toshev & Szegedy, 2014), scene labelling (Couprie et al., 2013), and a variety of other applications (Nithin & Sivakumar, 2015). CNNs are very close to ANNs, which may be seen as an acyclic graph with a well-organized collection of neurons. Unlike neural networks, CNNs only link the buried layers of neurons to the preceding layer, which contains a subset of neurons. This form of sparse connection allows the system to learn characteristics directly. Each portion of the CNN layer has two or more dimensional filters that are convolved with the layer's input. Deep convolutional networks yield hierarchical feature extraction (Aloysius & Geetha, 2018). CNN's architecture is made up of convolutional layers, pooling layers, and fully linked layers.

2.4.2.3 Artificial neural networks

ANN is a part of deep learning that is an advanced algorithm that is formed on the basis of the human brain. It is a blueprint to mimic the human brain. As our brain trains itself using previous data, the ANN may train through data and deliver answers in the sort of forecasts or classifications. ANN may also be known as a neural network that is made up of many nodes that are equivalent to the neurons as neurons work in the nervous system by exactly the same mechanism as nodes.

An ANN node is equivalent to neurons. The node has two parts. The first one is the summation part, and the second one is the activation function part. The summation part calculates the weighted sum of all the inputs ($x1\ w1\ +\ x2\ w2\ +\ -------+xn\ wn\ =\ \sum xi\ wi$).

Once the weighted sum is calculated, it is sent to the activation function. The activation function generates a particular output for a given node based on the input that is getting provided. Logistic regression, the statistical approach with which they have the most similarities, is a novel option for ANNs. Neural networks are algorithms that mimic the structure of the human brain (*How Neurl Networks Len Rom Eperience*, 1992). They are made up of a set of mathematical equations that are employed to replicate biological processes like learning and memory. For this purpose, neural networks have been constructed. Among other uses, they are used to diagnose acute myocardial infarction and acute pulmonary embolism and to forecast intensive care unit (ICU) resource utilization following cardiac surgery (Tu & Guerriere, 1992). The purpose of a neural network is identical to logistic regression modelling: to predict a result using the values of some predictor variables. The technique taken in constructing the model, on the other hand, is entirely different. Although there are several forms of neural networks, each node in the input layer is normally associated with each or every node in the hidden layer, so each node inside the hidden layer is generally linked to every node in the output layer (Tu, 1996). ANN is divided into three parts, as described in Figure 2.1

- Input layer
- Hidden layer
- Output layer

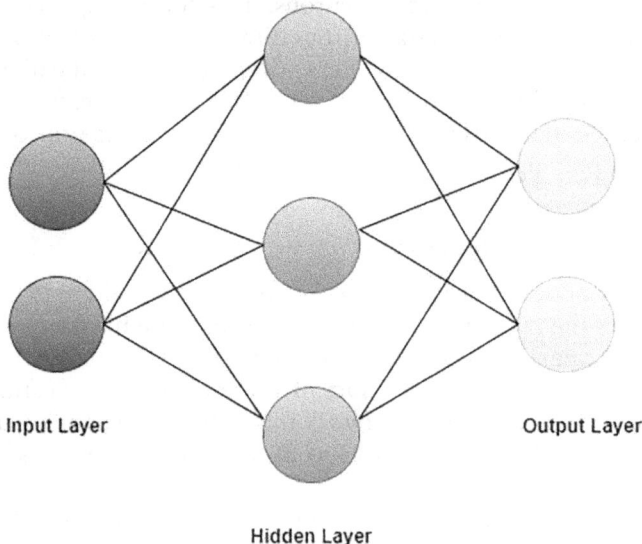

Figure 2.1 Architecture of artificial neural network.

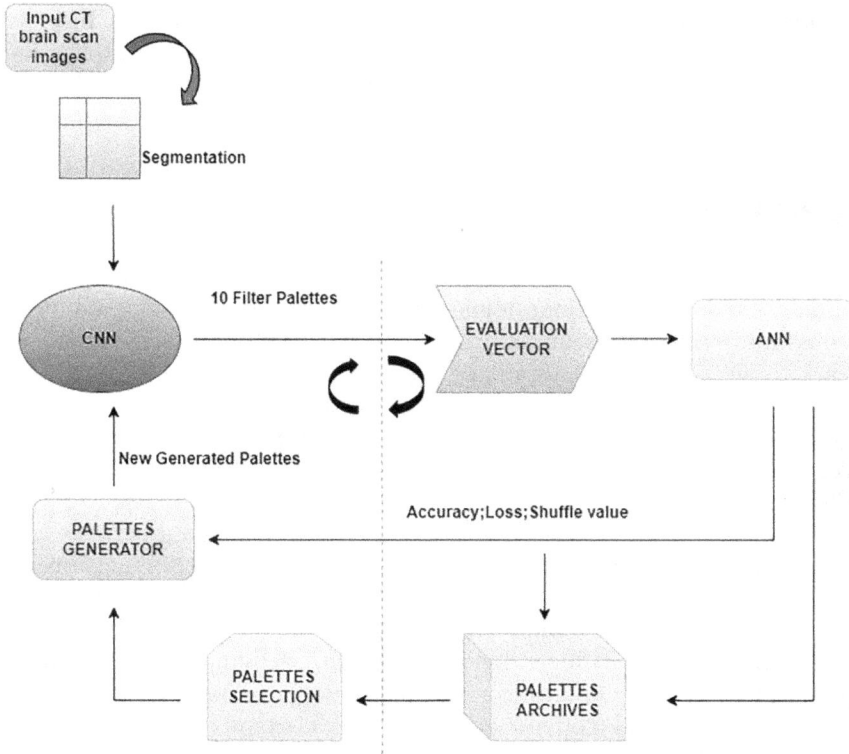

Figure 2.2 Correlation learning mechanism.

2.4.5 Considered learning algorithms

The suggested correlation learning mechanism (CLM) model is made up of CNNs that collaborate with traditional neural networks throughout the training stage (ANN). Both neuronal designs are part of a structure. It is learning how to analyze CT brain scans while sharing data in the form of filter palettes for CNN and numeric characterizing the evaluated picture for ANN. Figure 2.2 depicts the overall concept of CLM.

2.5 BEST PERFORMING ALGORITHM

Medical examinations in modern medical clinics are supported by computer technologies that employ computational intelligence to detect potential health concerns more efficiently. One of the most significant applications is the examination of CT brain images, where deep learning algorithms produce the most exact findings. In this paper, we present a unique CLM of DNN topologies that blends CNNs with the traditional design. The supporting neural network assists CNN in determining the best filers for pooling for convolution layers. As a result, the primary neural classifier learns more quickly and efficiently. The results reveal that our CLM model can achieve 96% accuracy, 95% precision, and 95% recall.

2.6 NORMALIZATION AND NEURAL CODING

The brain is built in modules. Engineers are well aware of the benefits of modularity as strong technologies are built on modules that could be copied and trumbled, such as transistors and web servers. This idea appears to be used by the brain in the following two ways *(i)* modular circuits and using *(ii)* modular calculations. Anatomic evidence shows the presence of canonical microcircuits, which are repeated across brain areas, such as the cerebral cortex (Rodney et al., 1991). Physiological or behavioral evidence points to the presence of canonical brain computations, which are typical computational modules that conduct the same core processes in diverse contexts. A canonical neural computation can use a variety of circuits and processes. Various brain areas or species may use different accessible components to execute it.

Exponentiation and linear filtering are two well-known paradigms of canonical neural computations. Exponentiation, a kind of thresholding, acts at the neuronal and network levels, for example, the mechanism through which eye and limb motions are produced. (Saito et al., 2018). This procedure serves several important functions, including preserving sensory selectivity, decorrelating signals, and establishing perceptual choice. A common computational method in sensory mechanism, which is linear filtering (weighted summing by linear receptive fields), is carried out, at least roughly, at several levels of vision, hearing, and somatosensation. It aids in the explanation of a wide range of perceptual events and may possibly be implicated in sensory and motor systems (Prasetyoputri, 2021).

In several neural networks, a third type of computation has been observed: divisive normalization. Normalization evaluates a ratio between the response of a single neuron and the overall activity of a clump of neurons. Normalization was proposed in the early 1990s to explain the non-linear characteristics of neurons in the primal visual cortex. Similar computations have previously been suggested to explain light adaptation of the retina, size invariance of the fly visual system, and associative memory in the hippocampus. Evidence gathered since then shows that normalization is involved in a wide range of modalities, brain areas, and animals.

Theorists have proposed numerous (not mutually incompatible) rationales for normalizing, the majority of which are connected to code efficiency and enhancing sensitivity. Normalization modifies the procurement of neural responses to make better use of the dynamic range offered, enhancing sensitivity to changes as input (Prasetyoputri, 2021). Light adaptation of the retina allows for great sensitivity to tiny changes in visual characteristics over a broad range of intensities. Normalizing reward values produces depiction capable of distinguishing between dollars and million dollars, hence expanding the reward system's effective dynamic range and the invariance with regard to some stimulus dimensions. Normalization in the fly's antennal lobe is hypothesized for allowing odorant identification and discrimination independent of concentration. In the retina where normalization discards information about the mean light level in order to retain invariant representations of other visual properties (e.g., contrast). Normalization in *V1* disposes contrast information used to encode picture patterns (e.g.,orientation), maximizing discriminability independent of contrast.

It is believed that MT encodes velocity independently of structural pattern. Normalization inside the ventral visual pathway might aid in the creation of object representations that are immune to changes (size, location, lighting, and occlusion) during the process of decoding a distributed neural representation. The responses of a group of neurons adjusted for distinct speeds and directions are assumed to represent visual motion in visual region MT. These responses may be seen as discrete samples of a likelihood density function for which the firing rate of each neuron is proportionate to a probability, the means of the distribution anticipates stimulus velocity and the variance of the distribution indicates the uncertainty in that prediction. If the firing rates are normalized to sum a constant, the mean & variance may be determined simply as weighted sums of the firing rates, the same constant for each stimulus. Differentiating between stimuli, normalization can help a linear classifier distinguish between neural representations of various inputs. A point in n-dimensional space represents the response of ˆ neurons to a stimulus. The points associated with comparable stimuli group together. A linear classifier distinguishes between stimulus categories by connecting them using hyperplanes. This is challenging if certain stimuli elicit indestructible reactions while others elicit weak responses; the hyperplane that establishes the boundary far from the origin might fail close to the origin, and vice versa. This difficulty is avoided via normalization.

Max-pooling (winner-take-all) (Katiyar, 2022) a neuronal population is normalized. It can work in two modes, i.e., when the inputs are almost equal, averaging them, and determining a winner, when one input is much higher than the rest, the competition is winner-take-all (max-pooling, choosing the maximum in inputs). Max-pooling is hypothesized to work across different brain systems and to underpin perceptual judgments by picking the neuronal subpopulation (or psychophysical channel) with the most responses. Object recognition models suggest numerous phases of max-pooling. Similar to the "biased competition" concept, attention may depend on normalization to change the calculation from average to max-pooling, thereby picking the sub-population with the most replies and suppressing the others.

When the outputs of a neural representation are duplicated, it is considered inefficient. Normalization can help to decrease redundancy. Normalization helps to ensure that *V1* reactions to natural visuals are statistically independent. As a result, the response *V1* summation field may be calculated using the responses of surrounding *V1* summation fields. Normalization removes the dependency, resulting in a more efficient representation. Similarly, in the fly olfactory bulb, reactions of the community of antennal lobe neurons (post-normalization) are more differentiable than olfactory system neurons (pre-normalization) (Carandini & Heeger, 2012).

2.7 CONCLUSION

There has been a significant renewed interest in neural modelling throughout the last decade. Work in this area has resulted in powerful new communication mechanisms that are being used for a growing number of medical conditions. The suggested CLM model learns quickly from data. We can see from the data that everyone has the potential to learn quickly and efficiently. The approach provides a fresh and simple concept of CNN composition. The palette might be made up of the number of filters to change the image and grids to extract the most significant aspects available. We picked those that produced the greatest results in brain tumor identification in our study experiments; however, the CLM can also be employed for other applications. The CLM may analyze numerous incoming palettes (parameter of the CNN architecture) from each cycle due to concurrent implementation.

REFERENCES

Abeles, M., Bergman, H., Gat, I., Meilijson, I., Seidemann, E., Tishby, N., & Vaadia, E. (1995). Cortical activity flips among quasi-stationary states. *Proceedings of the National Academy of Sciences of the United States of America*, 92(19), 8616–8620. 10.1073/pnas.92.19.8616

Ackermann, R. F., Finch, D. M., Babb, T. L., & Engel, J. (1984). Increased glucose metabolism during long-duration recurrent inhibition of hippocampal pyramidal cells. *Journal of Neuroscience*, 4(1), 251–264. 10.1523/jneurosci.04-01-00251.1984

Adey, W. R., Walter, D. O., & Hendrix, C. E. (1961). *Computer Techniques in Correlation and Spectral Analyses of Cerebral Slow Waves during Discriminative Behavior. 524*, 501–524.

Aloysius, N., & Geetha, M. (2018). A review on deep convolutional neural networks. *Proceedings of the 2017 IEEE International Conference on Communication and Signal Processing, ICCSP 2017, 2018-Janua* (November 2020), 588–592. 10.1109/ICCSP.2017.8286426

Brasil-Neto, J. P., McShane, L. M., Fuhr, P., Hallett, M., & Cohen, L. G. (1992). Topographic mapping of the human motor cortex with magnetic stimulation: factors affecting accuracy and reproducibility. *Electroencephalography and Clinical Neurophysiology/ Evoked Potentials*, 85(1), 9–16. 10.1016/0168-5597(92)90095-S

By, S. (2018). *Prediction of Asthma as Side Effect After Vaccination*. May.

Carandini, M., & Heeger, D. J. (2012). Normalization as a canonical neural computation. *Nature Reviews Neuroscience*, 13(1), 51–62. 10.1038/nrn3136

Couprie, C., Najman, L., & Lecun, Y. (2013). For Scene Labeling. *Pattern Analysis and Machine Intelligence, IEEE Transactions On*, 35(8), 1915–1929. 10.1109/TPAMI.2012.231

Cowan, N. (1998). Visual and auditory WM capacity. In *Trends in Cognitive Sciences* (Vol. 2, Issue 3, pp. 77–78).

Fan, J., Xu, W., Wu, Y., & Gong, Y. (2010). Human tracking using convolutional neural networks. *IEEE Transactions on Neural Networks*, 21(10), 1610–1623. 10.1109/TNN.2010.2066286

Gally, J. A., Montague, P. R., Reeke, G. N., & Edelman, G. M. (1990). The NO hypothesis: Possible effects of a short-lived, rapidly diffusible signal in the development and function of the nervous system. *Proceedings of the National Academy of Sciences of the United States of America*, 87(9), 3547–3551. 10.1073/pnas.87.9.3547

Gerstner, W., & Naud, R. (2009). How good are neuron models? *Science*, 326(5951), 379–380. 10.1126/science.1181936

Gevins, A., Smith, M. E., McEvoy, L. K., Leong, H., & Le, J. (1999). Electroencephalographic imaging of higher brain function. *Philosophical Transactions of the Royal Society B: Biological Sciences*, 354(1387), 1125–1134. 10.1098/rstb.1999.0468

Goodisman, M. A. D., & Asmussen, M. A. (1997). *Zyxwvutsrqp Zyxwvu Zyxwv Zyx Zyxwvuts Zyxwv Zyxwvutsrqponm Zyxwvutsr Zyxw Zyxwvutsrqp Zyxwvu Z Zyxwvutsrq Zyxwvu Zyxwvutsrq Zyxwvu*. 338, 321–338.

Herz, A. V. M., Gollisch, T., Machens, C. K., & Jaeger, D. (2006). Modeling single-neuron dynamics and computations: A balance of detail and abstraction. *Science*, 314(5796), 80–85. 10.1126/science.1127240

Hong, S., Ag, B., & Fairhall, A. L. (2007). *Single Neuron Computation: From Dynamical System to Feature Detector*. 3172, 3133–3172.

Horwitz, B., Friston, K. J., & Taylor, J. G. (2000). Neural modeling and functional brain imaging: An overview. *Neural Networks*, 13(8–9), 829–846. 10.1016/S0893-6080(00)00062-9

Horwitz, B., Rumsey, J. M., & Donohue, B. C. (1998). Functional connectivity of the angular gyrus in normal reading and dyslexia. *Proceedings of the National Academy of Sciences of the United States of America*, 95(15), 8939–8944. 10.1073/pnas.95.15.8939

Horwitz, B., & Sporns, O. (1994). Neural modeling and functional neuroimaging. *Human Brain Mapping*, 1(4), 269–283. 10.1002/hbm.460010405

Horwitz, B., Tagamets, M. A., & McIntosh, A. R. (1999). Neural modeling, functional brain imaging, and cognition. *Trends in Cognitive Sciences*, 3(3), 91–98. 10.1016/S1364-6613(99)01282-6

How Neural Networks Learn from Experience. (1992).

Jindra, R. H. (1976). Mass action in the nervous system. *Neuroscience*, 1(5), 423. 10.1016/0306-4522(76)90135-4

Katiyar, K. (2022). *AI-Based Predictive Analytics for Patients' Psychological Disorder BT - Predictive Analytics of Psychological Disorders in Healthcare: Data Analytics on Psychological Disorders* (M. Mittal & L. M. Goyal (Eds.); pp. 37–53). Springer Nature Singapore. 10.1007/978-981-19-1724-0_3

Katiyar, K., Kumari, P., & Srivastava, A. (2022). Interpretation of Biosignals and Application in Healthcare. In M. Mittal & G. Battineni (Eds.), *Information and Communication Technology (ICT) Frameworks in Telehealth* (pp. 209–229). Springer International Publishing. 10.1007/978-3-031-05049-7_13

Kwong, K. K., Belliveau, J. W., Chesler, D. A., Goldberg, I. E., Weisskoff, R. M., Poncelet, B. P., Kennedy, D. N., Hoppel, B. E., Cohen, M. S., Turner, R., Cheng -, H. M., Brady, T. J., & Rosen, B. R. (1992). Dynamic magnetic resonance imaging of human brain activity during primary sensory stimulation. *Proceedings of the National Academy of Sciences of the United States of America*, 89(12), 5675–5679. 10.1073/pnas.89.12.5675

Levitt, J. B., Kiper, D. C., & Movshon, J. A. (1994). Receptive fields and functional architecture of macaque V2. *Journal of Neurophysiology*, 71(6), 2517–2542. 10.1152/jn.1994.71.6.2517

McIntosh, A. R., & Gonzalez-Lima, F. (1991). Structural modeling of functional neural pathways mapped with 2-deoxyglucose: effects of acoustic startle habituation on the auditory system. *Brain Research*, 547(2), 295–302. 10.1016/0006-8993(91)90974-Z

Motion, O. F. (1990). *Differential Equations of Motion. March*, 66.

Nithin, D. K., & Sivakumar, P. B. (2015). Generic Feature Learning in Computer Vision. *Procedia Computer Science*, 58, 202–209. 10.1016/j.procs.2015.08.054

Ogawa, S., Menon, R. S., Tank, D. W., Kim, S. G., Merkle, H., Ellermann, J. M., & Ugurbil, K. (1993). Functional brain mapping by blood oxygenation level-dependent contrast magnetic resonance imaging. A comparison of signal characteristics with a biophysical model. *Biophysical Journal*, 64(3), 803–812. 10.1016/S0006-3495(93)81441-3

Oxide, T. H. E. N., & Of, D. (1947). *The Nitrous Oxide Method for the Quantitative. 1*, 476–483.

Pineda, F. J. (1988). Dynamics and architecture for neural computation. *Journal of Complexity*, 4(3), 216–245. 10.1016/0885-064X(88)90021-0

Pineda, F. J. (1989). *Recurrent Backpropagation and the Dynamical Approach to Adaptive Neural Computation. 172*(1986), 161–172.

Poli, R., Cagnoni, S., Coppini, G., & Valli, G. (1991). A Neural Network Expert System for Diagnosing and Treating Hypertension. *Computer*, 24(3), 64–71. 10.1109/2.73514

Prasetyoputri, A. (2021). Detection of Bacterial Coinfection in COVID-19 Patients Is a Missing Piece of the Puzzle in the COVID-19 Management in Indonesia. *ACS Infectious Diseases*, 7(2), 203–205. 10.1021/acsinfecdis.1c00006

Reggia, J. A. (1993). *Neural Computation in Medicine. 5*, 143–157.

Rodney, B. Y., Douglas, J., & Martin, K. A. C. (1991). *Physiology*, 440, 735–769.

Saito, B., Nakashima, H., Abe, M., Murai, S., Baba, Y., Arai, N., Kawaguchi, Y., Fujiwara, S., Kabasawa, N., Tsukamoto, H., Uto, Y., Ariizumi, H., Yanagisawa, K., Hattori, N., Harada, H., & Nakamaki, T. (2018). Efficacy of palonosetron to prevent delayed nausea and vomiting in non-Hodgkin's lymphoma patients undergoing repeated cycles of the CHOP regimen. *Supportive Care in Cancer*, 26(1), 269–274. 10.1007/s00520-017-3845-y

Seemanthini, K., & Manjunath, S. S. (2021). Recognition of trivial humanoid group event using clustering and higher order local auto-correlation techniques. In *Cognitive Computing for Human-Robot Interaction: Principles and Practices* (Issue August). 10.1016/B978-0-323-85769-7.00001-X

Sincich, L. C., Horton, J. C., & Sharpee, T. O. (2009). Preserving information in neural transmission. *Journal of Neuroscience*, 29(19), 6207–6216. 10.1523/JNEUROSCI.3701-08.2009

Sokoloff, L., Reivich, M., Kennedy, C., Rosiers, M. H. D., Patlak, C. S., Pettigrew, K. D., Sakurada, O., & Shinohara, M. (1977). the [14C] Deoxyglucose Method for the Measurement of Local Cerebral Glucose Utilization: Theory, Procedure, and Normal Values in the Conscious and Anesthetized Albino Rat. *Journal of Neurochemistry*, 28(5), 897–916. 10.1111/j.1471-4159.1977.tb10649.x

Srivastava, A., Seth, A., & Katiyar, K. (2021). Microrobots and Nanorobots in the Refinement of Modern Healthcare Practices. *Robotic Technologies in Biomedical and Healthcare Engineering*, May 2021, 13–37. 10.1201/9781003112273-2

Toshev, A., & Szegedy, C. (2014). DeepPose_Human_Pose_2014_CVPR_paper.pdf. *2014 IEEE Conference on Computer Vision and Pattern Recognition*, 1653–1660. https://www.cv-foundation.org/openaccess/content_cvpr_2014/html/Toshev_DeepPose_Human_Pose_2014_CVPR_paper.html

Tu, J. V., & Guerriere, M. R. (1992). Use of a neural network as a predictive instrument for length of stay in the intensive care unit following cardiac surgery. *Proceedings / the ... ... Annual Symposium on Computer Application [Sic] in Medical Care. Symposium on Computer Applications in Medical Care*, 666–672.

Tu, J. V. (1996). *Advantages and Disadvantages of Using Artificial Neural Networks versus Logistic Regression for Predicting Medical Outcomes.* 49(11), 1225–1231.

Chapter 3

Neural networks and neurodiversity
The key foundation for neuroscience

Hera Fatma, Harshit Mishra, and Kalpana Katiyar
Department of Biotechnology, Dr. Ambedkar Institute of Technology
for Handicapped, Kanpur, Uttar Pradesh, India

3.1 INTRODUCTION

Neuroscience, which includes neurology, psychology, and biology, is a relatively recent field of study. Over the past century, significant progress has been made in the understanding that covers the anatomical features, physiology, biochemistry, and pharmacology of the mammalian brain (Parra, Hruby, and Hruby, n.d.). Fundamentally, brains are organic computer devices. They enable the capacity of an organism to see and comprehend its environment, combine and decide from several data streams, and adjust to a changing situation by turning a flood of complicated and confusing sensory input into coherent cognition and action. In light of this, it is probably not surprising that biology has long served as an inspiration for computer science, the field of creating artificial computing systems (Cox and Dean 2014a). Although there are many prospects for collaboration between computer science and neuroscience, the path to developing neurologically algorithms has been difficult and winding. Here, we look back at the historical relationships between computer science and neuroscience and ahead to a new era of potential cooperation enabled by recent rapid developments in experimental neuroscience and biochemically based computer vision approaches. Our attention is on the applications of neuroscientific algorithms, areas that have been successful, where they are currently ineffective, and where deeper links are expected to be beneficial.

Autistic and other neurodivergent advocates and activists have historically led and made up the majority of the neurodiversity movement, with limited engagement from neurotypical stakeholders. We are starting to notice a favorable change in neurotypical stakeholders' views about autism as the neurodiversity movement gets support within the larger autistic community. Treatment objectives are increasingly centered on problems of critical concern for the autistic community rather than the normalization of autistic persons, and strengths-based methods to support and intervention are widely acknowledged as best practices (den Houting 2019). Anatomical, electrophysiological, and computational restrictions are taken

 DOI: 10.1201/9781003398066-3

into account when using recurrent neural networks (RNNs), a family of computer models, to explain neurobiological events (Barak 2017). RNNs may be taught using input-output examples or they can be created to implement a specific dynamical principle. Making use of trained RNNs for computing jobs and as neural phenomenon elucidations have advanced significantly in recent years. I'll go through how using trained RNNs in conjunction with the use of reverse engineering might offer an alternate structure for modelling in neurological science and perhaps act as a potent tool for developing hypotheses. Despite recent developments and potential advantages, there are still several basic holes in the theory of these networks (Srivastava and Jha 2022). We will talk about these difficulties and potential solutions.

3.2 WHAT IS NEUROSCIENCE?

Neuroscience refers to a relatively recent field that integrates biology, psychology, and neurology. The understanding of biochemistry, psychology, pharmacy, and structural features of the mammalian brain has advanced significantly during the past century. Considering that cognitive neurosciences have developed, which concentrate explicitly on comprehending greater-level cognitive functions using imaging technologies, comprehension of some of the basic emotional, mnemonic, attentional, perceptual, and cognitive processes has also improved (Parra, Hruby, and Hruby, n.d.). We now have a better grasp of the extremely complicated processes that underlie arithmetic, reading, reading comprehension, and speaking and articulation thanks to neuroimaging. Thereafter, it seems appropriate thinking about how we may apply improved knowledge of our mammalian brain growth and activity to research pedagogic issues.

According to its broad definition, neuroscience studies how the brain develops and remembers at all levels of organization, from molecules and cells to whole brain systems (like the network of neural regions and pathways that underlies our capacity for language comprehension and speech, for example). This emphasis on memory and learning can occur at different extents. Studying synaptic processes and cell signalling – where one neuron joins another neuron through synapse – is critical for comprehending knowledge, but is looking closely at the activities of particular areas in the brain, such as the hippocampus, using invasive or non-invasive techniques (Material 2002).

Our vocabulary for describing the brain and our grasp of modern man-made technology have been crucial to our understanding of the brain throughout history. Descartes used hydraulic analogy and the flow of fluids to illustrate how the mind works (Cox and Dean 2014b). More and more, "channels" and "frequencies" were used to describe how the brain functions, throughout the radio era (Srivastava, Seth, and Katiyar 2021).

Perhaps unsurprisingly, we speak in the language of contemporary technology today. Neuroscientists talk more and more about the "circuits" and "computations" that occur within neurons and how different parts of the brain interact to create "networks" of activity. It would be easy to dismiss the "computational" perspective on neurological science as just another fad, but it is an analogy with deeper roots: beyond the adage "the brain is a computer," machine learning science offers a strict systematic approach and methodologies for reasoning about information systems, isolating what gets tabulated (the "algorithm") from the direction it gets tabulated (the "implementation") (Ogden and Miller 1966). Today's world has access to a tremendous amount of computing power. Since the invention of the Internet, we have become accustomed to interacting with enormous computer networks and have developed the technology to take advantage of their combined capacity. Several organizations have started massive, international projects to model individual brain cells or whole brains in silico (McCalpin 1995). General perception and object identification in specific, however, presents a fascinating case study for the confluence of computers and neurological science, even if a thorough examination of all relationships between computers and neurological science is outside the purview of the current chapter. Here, we examine this interface's past and present and offer potential channels for further cross-pollination (Cox and Dean 2014b).

3.3 ARTIFICIAL NEURAL NETWORK: A BRIEF CHRONOLOGY

Unexpectedly far back in the timeline of computers lies the origin of biologically inspired algorithms. In 1943, McCulloch and Pitts established the concept of a "integrate and fire" neuron, and Hebb was the first to put out the notion of assimilation and accommodation in the brain cells in the late 1940s: "What fires together, wires together" (Graben and Wright 2011). In contrast, the transistor was not created until 1947, usable integrated circuits did not arise until the end of late 1950s, "minicomputers" and mainframes didn't become widely used until the late 1960s, and personalized computer systems didn't come into being until the late 1980s and 1990s. The fact that the proposal would come to pass before actual application by such a long time is evidence of these early pioneers' vision (Cox and Dean 2014b). Rosenblatt's "perceptron," which presented a straightforward configuration of neurons with output and input that may judge based on input vectors, was one of first examples of a neural network that can learn. Since the original perceptron could only learn linear functions of the inputs, it was found to be fundamentally limited. As a result, neural network research temporarily suffered by the opposing "symbolic artificial intelligence" faction, which sought to mimic intelligence through processes that used abstract representations rather than deriving inspiration directly from the mammalian brain's neural network architecture (White and Rosenblatt 1963). The conceptual limitations of the perceptron were overcome by the inclusion of non-linear between the output and input of the networks and a layer of units with activation functions, leading to the emergence of numerous types of artificial neural networks over the following two decades (Thibault and Grandjean 1991).

Through the 1980s, artificial neural networks proliferated, and hope was high. The study of many early neural network types with the goal of addressing a variety of issues, from vision to language, became known as "connectionism," and the name quickly gained popularity. The condition in neural networks was significantly advanced by a number of researchers (including LeCun, Bengio, Hinton, and Schmidthuber, to mention a few), and they were progressively used to address a range of problems in the real world. A potent tool for image analysis is the convolutional neural network, which made significant contributions to the developing science of computer vision by doing very well on the challenge of hand-typed digit identification, a practical use of neural networks. In the meantime, the creation of neural networks has been guided by neurology, which has inspired the architectural aspects of neural networks (e.g., pooling in Fukushima's noncognition water from basic to complicated) (Cox and Dean 2014a).

3.3.1 Do deep learning and neuroscience still need each other?

Autistic and some other neurodivergent advocates and activists have historically been at the forefront of the neurodiversity movement, with neurotypical stakeholders playing a far smaller role. Now that the neurodiversity movement is gaining support throughout the larger autism community, we are starting to notice a favorable change in neurotypical stakeholders' views about autism (den Houting 2019). The fact that there were difficult or unavailable ways to teach a machine using numerous levels has been a main drawback of classic artificial neural networks. However, it is clear from a visual neuroscience standpoint why building multilayer networks is appealing. The ventral visual pathway in primates is structured as a hierarchical network of interconnected visual regions and is hypothesized to provide visual shape and object vision (Cox and Dean 2014b). "Is neuroscience still relevant to machine learning with the current excitement around contemporary deep learning techniques?" is a natural question to pose. There is no doubt that deep learning techniques have made significant progress, and a lot of optimism that deep learning techniques may still be continued to solve a number of the machine learning issues that are now plaguing the field.

The transfer of concepts from computational sciences to neurological sciences, meanwhile, has been irregular and hasn't resulted in the best advancement. In light of this, it could seem simple to conclude that neuroscience is no longer necessary for machine learning. In fact, A.I. Spring's major player Yann LeCun was recently quoted as noting that while biology might serve as inspiration, we shouldn't let it lead us astray. Out-of-set generalization is a concern for contemporary computer vision. Performance evaluation is a key difficulty in computational perspective. Performance in computational perspective is frequently measured in comparison to data sets (K. Katiyar 2022). Torralba and Efros, however, have neatly demonstrated the majority of the system learned on one data set outperform those trained; a prejudice in such data sets may be seen when one is compared to another that has the same item categories (Hyvärinen 2010). Deep learning literature also lists additional, less obvious warning indicators of problems. Szegendy and colleagues, for instance, demonstrated how adding purposefully created "noise" to photos might result in images being randomly misclassified by a con-temporary deep learning system (Szegedy et al. 2014).

3.4 NEURO-IMAGING METHODS FOR COGNITIVE DEVELOPMENTAL NEUROSCIENCES

Researches on neuroimaging are predicated on the idea that every intel-lectual endeavor places certain demands on the brain, and that these demands will be fulfilled by alterations to brain activity. Through PET (positron emission tomography), local blood flow may be measured explicitly or implicitly (by these changes in activity) (fMRI). ERPs can monitor dynamic interactions between mental processes (Parra, Hruby, and Hruby, n.d.). Children should not undergo PET, which requires infusion of radioactive tracing elements (tracers). The tracers are present in greater quantities in brain regions with increased blood flow, allowing images of the radiation distribution to be formed and enabling the localization of various neuronal activities. The localization of brain activity is also made possible by fMRI (functional magnetic resonance imaging). The participant must be placed within a sizable magnet (e.g., a huge tube) for this approach to operate. It measures the computed tomography signal produced by the proton of water molecules in brain cells (Geake and Cooper 2003).

Most fMRI studies use the changes in BOLD response as their primary outcome indicator. Participants receive headphones to protect their ears from the loud noise inside the magnet as well as a panic alarm switch, i.e., the magnet is claustrophobic. Considerations have made it difficult to modify fMRI for usage with minors (who are frequently moving as well, reducing imaging precision). Moreover, the number of such investigations is increasing due to the development of specifically designed coils and less confining head scanners (Parra, Hruby, and Hruby, n.d.). The first few years of life see a decline in ERP latencies (concurrent with myelinization), and late in childhood they reach adult levels. The temporal course of brain processing has been extensively documented by ERP investigations, which are also sensitive to millisecond changes (S. Katiyar and Katiyar 2021). To comprehend the underlying cognitive processes, one uses the sequence of recorded potentials, together with their magnitude and duration.

3.5 NEUROMYTHS

The fascinating phrase "neuromyths," which was used in the OECD study on studying the mammalian brain (OECD, 2002), illustrates how quickly as well as easily scientific material may be misrepresented as having educational value. The three myths that the OECD report focuses on the most are the widespread idea that hemispheres differ (e.g., learning using the left or right brain), the idea that certain types of education must take place during "critical periods" when the brain is most plastic, and the notion that the best educational interventions should coincide with synaptogenesis (Berns, Cohen, and Mintun, n.d.) With regard to neuro-myth, the statements about the left and right brains are likely supported by the fact that various talents are localized in distinct hemispheres, which is a form of hemispheric specialization. Many parts of analyzing language, for instance, are left-lateralized (nevertheless, given what we have shown, in blind persons or among those who go abroad to different language group later in infancy). The right hemisphere is lateralized in several elements of facial recognition, in contrast. However, the normal brain has a significant number of cross-hemispheric connections, and in every cognitive activity that has been studied thus far using neuroimaging, including language and facial recognition tests, both hemispheres collaborate (Caspi et al. 2002).

3.6 NEURAL NETWORKS

A highly parallel distributed processor called a neural network is described as having a natural predisposition to retain experience information and make it accessible to users. Neural networks are made up of basic processing units. In a wide variety of applications, including robotics, speech recognition, face identification, medical applications, manufacturing, and economics, many researchers employ neural networks (Abdel-Nasser Sharkawy 2020). Deep network researchers started to amass a constant stream of useful accomplishments. Fully-connected designs, which examine a picture with a number of filters at each level of a high hierarchy, have shown to be effective at very successful at this time in the field of vision. The spatial stationarity in natural images – where a group of comparable visual elements tends to exist at various spatial places in an image – is naturally

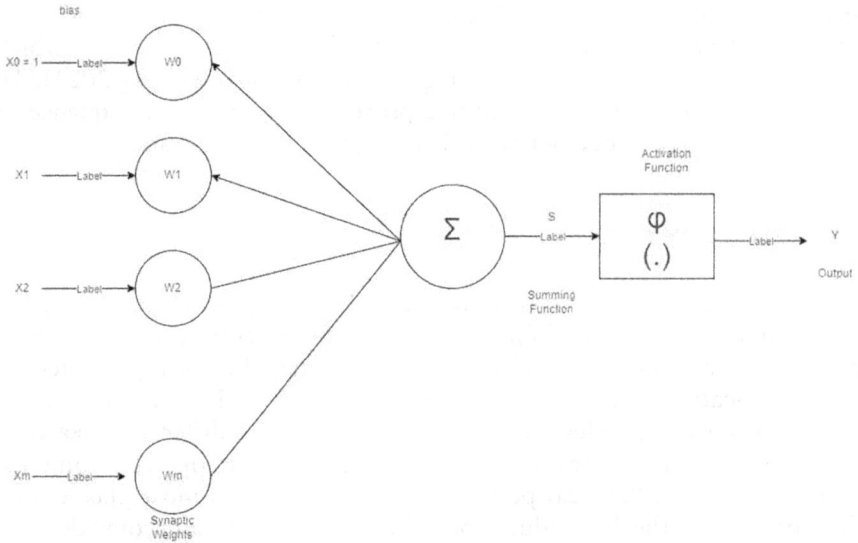

Figure 3.1 Non-linear neuron model.

captured by these systems, which require less weight in training, and perform effectively even if there are no weights learned (Cox and Dean 2014a). Numerous medical uses and diagnoses employ neural networks (Abdel-Nasser Sharkawy 2020).

3.6.1 Neuron models

The information-processing unit, or neuron, is essential to the functioning of a neural network. Figure 3.1: *Non-linear Neuron Model* presents the block diagram illustrating the neuron concept. The design of a family of massive neural networks is based on this figure (Barak 2017).

3.6.2 General properties of neural networks

The numerous benefits that neural networks offer make them popular in a wide range of applications. The following is a summary of these benefits.

- A neural network is an effective method for identifying non-linear systems (Sharkawy, Koustoumpardis, and Aspragathos 2020).
- The massively parallel distributed architecture of the neural network and its capacity for learning and generalization give it tremendous computational power. Given the right training data, it was demonstrated that such a neural network might approximate any complicated (large-scale) linear or non-linear function ("Fundamentals of Machine Learning and Softcomputing" 2006).

- The neural network is essential for identifying dynamic systems and detecting faults since it not only may be used to identify when a problem occurs but also gives a post-fault model of the robotic manipulator. If practicle, the failure can be accommodated using this post-fault model, which can also be utilized to isolate and diagnose the issue (Wang, Czerminski, and Jamieson 2021).
- The estimate of seamless batch data comprising the input, output, or perhaps gradients data of a function, as well as the approximation of a function's derivatives, are two examples of how neural network has some universality (Abdelhameed and Tolbah 2002).

3.6.3 Neural network classification

Here, we focus on the three primary neural networks —MLFNN, SLFFNN and RNN.

3.6.3.1 Multilayer feedforward neural network (MLFFNN)

In artificial feedforward neural networks, there isn't a recurrence in the links in between units. The first artificial neural networks were referred to as feedforward neural networks. They are also less complicated than RNNs, their counterpart. They were referred to as feedforward since information can only advance via the network's input nodes, hidden nodes (if any are present), and output nodes in that order (no loops) (Abdel-Nasser Sharkawy 2020).

3.6.3.2 Single-layer feedforward neural network (SLFFNN)

The network output layer of the computing nodes is referred to as the single layer (neurons). Source nodes' input layer is not counted since no computation takes place there, as depited in Figure 3.2: *Single-Layer Feedforward Neural Network.*

3.6.3.3 Recurrent neural network (RNN)

The neural models known as RNNs are those that can incorporate context into their decision-making process. A feedforward neural network differs from an RNN in that it contains at least one feedback loop (Wang, Czerminski, and Jamieson 2021). The Elman and Jordan RNN models are the two primary variations of the RNN, sometimes known as "simple" RNNs, that have been suggested in the literature. An input layer, a hidden layer, a delay layer, and an output layer are layers that make up both the Elman and Jordan neural networks. While the Jordan neural networks' delay neurons are supplied from the output layer, the Elman neural networks' delay neurons are fed from the hidden layer (Abdel-Nasser Sharkawy 2020).

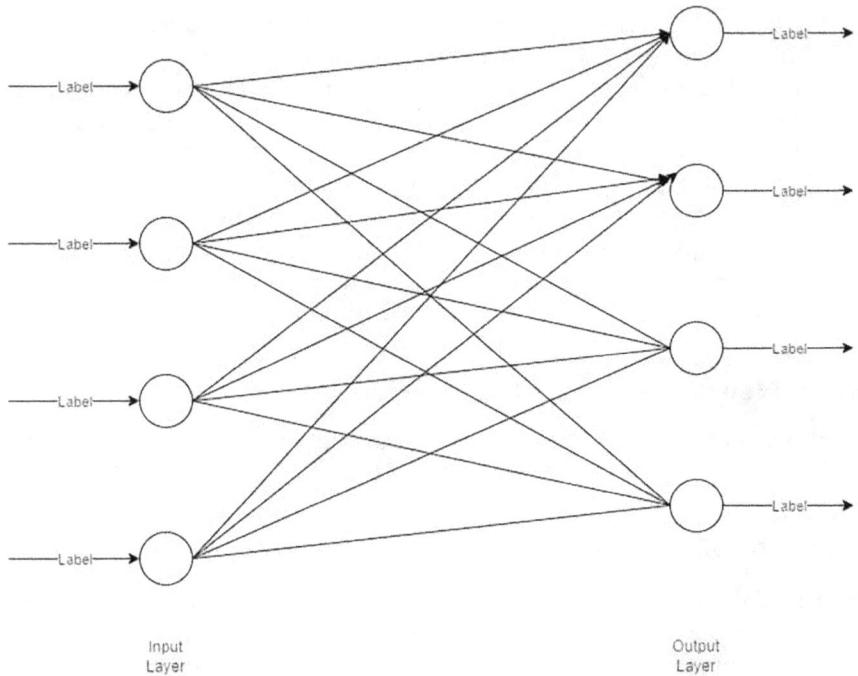

Figure 3.2 Single-layer feedforward neural network.

3.7 RNNS AS A TOOL OF NEUROLOGICAL SCIENCE RESEARCH

Continuous arrangements of enhancement issues are frequently required in logical and design issues, including signal handling, system verification, channel configuration, capability estimation, and regression analysis, and neural networks have been generally explored for this reason. The quantities of decision variables and limitations are typically extremely enormous, and wide scope advancement methodology are considerably more challenging when they must be finished progressively to upgrade the output of a dynamical framework. For such applications, classical enhancement strategies may not be sufficient due to the issue of dimensionality and tough prerequisites on computational time (K. Katiyar, Kumari, and Srivastava 2022). The neural network approach can tackle enhancement issues in running times significant degrees quicker than the most famous enhancement calculations executed on universally useful advanced computer systems.

RNNs are a gathering of computational models that are generally carried out as a device to portray neurobiological cycles taking into consideration the electronic, physiological, and computational entanglements (Barak 2017). The excellent spotlight on innovative work on these

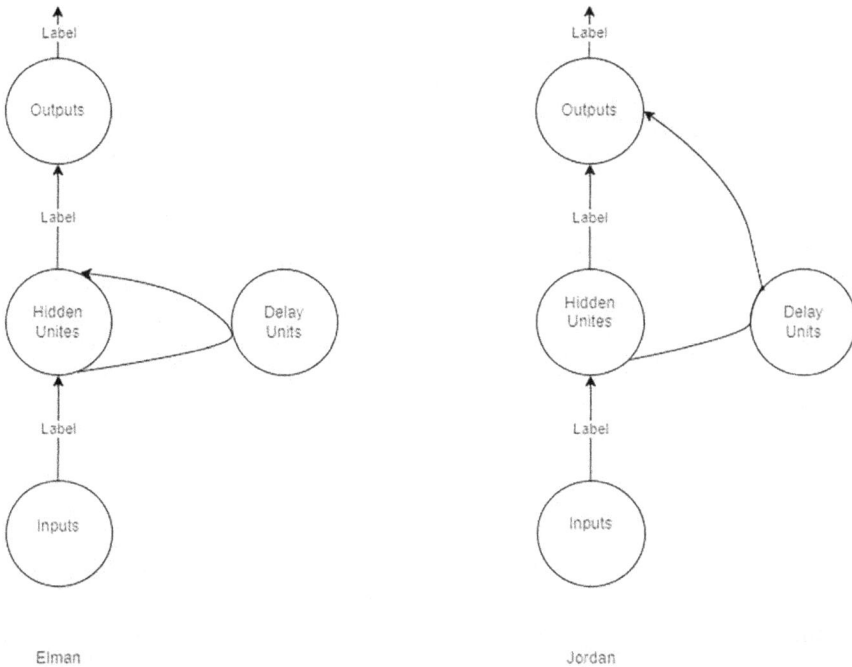

Figure 3.3 Types of RNNs.

neural networks has been used during the 1990s. They have been intended to comprehend time shifting or sequential patterns. A recurrent net is a kind of neural network which includes feedback (or closed loop connections). A few examples of RNN incorporate Hopfield, Boltzmann machine, BAM, and so forth. RNN strategies have been generally executed to a wide scope of issues. Basic to some degree, RNNs were created in the 1980s with the intent to learn strings of characters. RNNs have additionally resolved the issues that incorporate dynamical frameworks with time sequences of events. The two primary variations of RNN, likewise called "simple" RNNs, are, the Elman and the Jordan RNN models. Both the Elman and Jordan neural networks comprise a delay layer along with an input, hidden, and output layer. The postpone neurons of an Elman neural network are taken care of from the hidden layer, while the defer neurons of a Jordan network are taken care of from the output layer (Abdel-Nasser Sharkawy 2020). Figure 3.3: *Types of RNNs* below depicts the different types of RNNs.

3.7.1 RNNs as an important model for computations

RNNs are a type of computer models that are employed for problem solving as well as for the explanation of neurobiological phenomena and dealing issues in machine learning (Sussillo 2014). These networks lack a clear definition of upstream or downstream and allow a neuron to provide information to any other neuron in the network. As a result, the network's present state as well as the current stimulus has an impact on the activity of its neurons. Such networks are perfectly suited for computations that take place over time, including keeping information in working memory or gathering evidence to support a choice, because of this characteristic. The justification for adopting RNNs as model comes from both electro-physiological and anatomical considerations. A substantial portion of the output from practically all cortical regions is directed toward the region of origin. The normal cortical network is hence continually linked to itself. Additionally, an RNN has the capacity to produce complex intrinsic activity patterns that are similar to the continuing activity seen in the brain (Barak 2017).

3.7.2 RNNs designing

Designing the connections based on intuition is one strategy that produced numerous significant results. Take the Romo lab's delayed discrimination task as an illustration. In this task, a monkey must decide whether the frequency of the first vibrotactile stimulus is higher or lower after receiving two stimuli separated by a short period of time greater in contrast to the second one. As a first stage in the modelling process, we idealize the task by stating that the behavior is essentially an input-output transformation. The output transiently increases only if the first input amplitude is greater than the second one, and decreases otherwise. The input is made up of two pulses whose amplitudes correspond to the vibration frequency in the experiment. The realization that a representation of the first frequency that does not alter over time is a useful element for completing this job is now the insightful stage (observing the neural activity can also yield insights). The two stimuli could be compared using such a representation: Figure 3.4: *RNN with Hidden Neurons*.

The network should represent an abstract variable called f1 that, in the terminology of dynamical systems, follows the following low-dimensional dynamics during the delay: df1/dt = 0. Using intuition or computational techniques, these dynamics can be converted to the connectivity of an RNN (Machens, Romo, and Brody 2005). In the N-dimensional phase space, every value of the variable f1 corresponds to a certain point x. These points are gathered into a line attractor, which was designed into the network dynamics as a one-dimensional manifold of fixed points. Different tasks result in various dynamical objects, such as saddle points that mediate judgments, line attractors that implement the accumulation of evidence, point attractors that represent associative memories, and many more (Hopfield 1982). A low-dimensions dynamical system is used

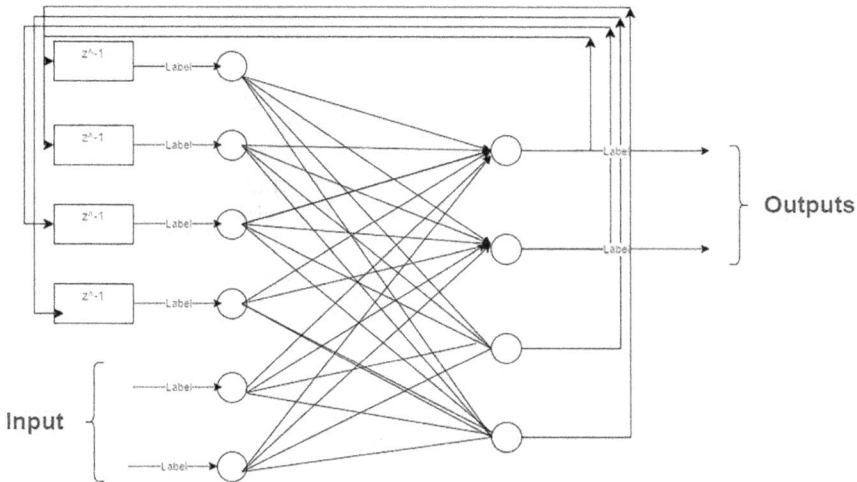

Figure 3.4 RNN with hidden neurons.

in each of these instances to generate a hypothesis about the underlying mechanism, which is then translated into a high-dimensional RNN. Because it makes all assumptions clear, the implementation serves as a consistency check for the hypothesis (Abbott 2008). Additionally, predictions may be made by comparing the activity of model neurons to that observed experimentally.

3.7.3 Functionality and optimization

Various utilitarian roles have been proposed for recurrent associations, by both the neuroscience and computer vision networks. One normal thought is that repetition empowers relevant data to be consolidated to improve generally equivocal sources of information. One can perceive profoundly debased images when an outsourced setting furnishes additional hints. Consolidating such a setting, for example in a Bayesian system, is a famous idea, however, one that still needs to be figured out. Likewise, various models place explicit jobs for hierarchical criticism associations in distributing consideration regarding various pieces of a scene. New apparatuses in neuroscience progressively give test access to straightforwardly concentrate on these connections. For example, infections presently exist that can bounce across a solitary neural connection, conveying hereditarily encoded markers and opsins that empower the action of neurons that give a contribution to a given objective to be estimated and controlled. RNN approaches give a clear idea of these connections, and their functions will be one region where neuroscience and computer vision could appreciate exceptional cooperative energy (Cox and Dean 2014b).

When a network is optimized, it is told what it should achieve but is not given many specific instructions on how to do it. As a result, optimizing and evaluating RNNs may be used to generate hypotheses for the next experiments and data analysis if the processes used by RNNs after optimization can be understood. For instance, the network could find a totally unexpected solution to the issue. It's possible that the best (or locally optimal) solution produced by optimization will take into account insignificant elements of the issue but nonetheless have an impact and alter the underlying answer. Modelling gain fields in the parietal cortex, as well as pattern production and neural dynamics in the motor system, are early examples of approaches that use this strategy (Sussillo 2014).

3.8 RNNS CAN BE TRAINED WITHOUT INTUITION

A different approach to building functioning RNNs is training, which is widely utilized in the machine learning community. The computational strength of RNNs comes from the fact that a neuron's activity is influenced by more than just the current flow, not only by the network's input but also by the network's state, which keeps track of previous inputs. This benefit comes with a cost because RNNs are challenging to train. However, substantial advancements in training techniques recently have made it possible to use RNNs for a range of tasks (Jaeger and Haas 2004). Consider the above-mentioned vibrotactile discrimination task. By providing the network with a large number of example trials and adjusting connectivity in accordance with a learning rule, the connectivity can be taught.

The outcome will be a high-dimensional neural network that can solve the problem, and it will be possible to compare the activity of the model neurons to that observed experimentally (Enel et al. 2016). In some instances, the trained example compared to designed networks has a higher correspondence to genuine neurons (Barak et al. 2013).

The final trained network, in contrast to the intended networks, resembles a mysterious black box. The fundamental method through which the network can complete the task is unknown to us. It appears that by skipping the "intelligence hypothesis" stage, we have replaced the poorly known brain with yet another poorly understood complex system (RNN). But as we'll see below, this substitute is a lot more easily accessible for study (Gao and Ganguli 2015).

3.9 HYPOTHESIS AND THEORY GENERATION

If the learned artificial network can be successfully reverse engineered, we will have inadvertently uncovered a hypothesis rather than deliberately considering it (Sussillo 2014). This was the situation in the context-dependent accumulation of evidence, which led to the discovery of a pair of line attractors with a specific relationship between their left- and right-eigenvectors (Mante et al. 2013). Although the finding of this mechanism was partially an automation of the scientific method, it was not hypothesized. This is because determining how much reverse engineering is actually interpretable is a judgment call. The mechanism discovered for the vibrotactile discrimination example was both anticipated and unexpected. On one hand, as was predicted for such a task, requiring varied delay times resulted in the creation of a line attractor. On the other hand, it was discovered that the decision itself was mediated on some networks by a saddle point and on other networks by a progressive distortion of the output pulse.

The most recent findings and understandings from training RNNs for different tasks are encouraging. However, there is a significant knowledge gap between what we know about designed or random networks and what we know about such networks (Hopfield 1982). What are the restrictions of this strategy? Where will the networks fail, and which tasks can be trained? The solutions discovered by training algorithms are how invariant? What aspects of the single neuron model being utilized or the learning algorithm itself affect the solutions? Modern theories for trained RNNs must be developed in order to address these and numerous other unresolved concerns. Such theories have a variety of directions. In terms of mathematics, it is feasible to train several networks on the same job while systematically changing particular network architecture, for instance. This strategy was refined in a recent study by training a network on a task with many phases (Enel et al. 2016). The researchers demonstrated that creating an explicit representation of the task phase, as opposed to letting one emerge implicitly during training, enhanced performance and offered a superior fit to electrophysiological data from monkeys executing this task. Utilizing straightforward tasks that lend themselves to more in-depth study and serve as building blocks for more complicated environments is another strategy. Given that many trained networks have been found to contain fixed points, examining networks that have been trained to contain a number of specified fixed points is a logical first step in this direction. Tools from mean field theory and systems theory were recently used to produce a better match to the monkeys' electrophysiological data collected during this task. Utilizing straight forward tasks that lend themselves to more in-depth study and serve as building blocks for more complicated environments is another strategy. Given that many trained networks have been found to contain fixed points, examining networks that have been trained to contain a number of specified fixed points is a logical first step in this direction. The dynamics of such networks were recently analyzed using a mix of mean field and systems theory methods (Rivkind and Barak 2017).

3.10 INTRODUCTION TO NEURODIVERSITY

Unfortunately, it is difficult to define "neurodiversity." The neurodiversity can be segregated into three distinct meanings. First and foremost, "neurodiversity" might simply refer to the fact that people have a variety of minds and brains, just as "biodiversity" can refer to the fact that living things are various. In this sense, since no two people have exactly the same mind or brain, even groups of neurotypical people are neurodiverse. It is far more difficult to describe "neurodiversity" in its various contexts. Although some (Kapp 2020) prefer the term "neurodiversity framework". The "neurodiversity movement," refers to an activist group that works to enhance the rights and welfare of persons with neurologically abnormal disabilities, or "neurodivergent" people. This concept is very different from the term "neurodiversity approaches".

In many aspects, the neurodiversity methods do actually resemble a paradigm. Both concepts are hard to describe; therefore, it makes sense to think that individuals rely mainly on their everyday assessments of whether something adheres to a neurodiversity approach on how closely it resembles other examples that have been shown to do so.

The neurodiversity approaches aim to prescribe a proper course of action in relation to human neurocognitive diversity, much like paradigms attempt to prescribe a proper course of action for conducting science. Furthermore, they are largely action-oriented and prescriptive, again similar to paradigms. However, compared to a scientific paradigm, the neurodiversity methods seem far more varied. They are not just difficult to describe; as will be covered in more detail below, different people seem to have fundamentally differing opinions on key aspects of what a proper neurodiversity strategy should entail. It could be more accurate to refer to different "neurodiversity approaches" rather than a single "neurodiversity paradigm" (Dwyer 2022).

3.11 NEURODIVERSITY: THE SITUATION OF INCLUDING AUTISTIC EMPLOYEES AT WORK

Rapid progress of technology over the past few years has resulted in substantial changes in workplace employment practices due to greater computing capacity, the massive growth of technical data and massive algorithmic advancements. The previous technological revolutions have been primarily brought about by improvements in popularly used technologies, such as steam power, electricity, and computer-based technologies. However, the current Industrial Revolution includes a paradigm shift across all fields of study, industries and economies because it is raising significant political and philosophical issues in the process (Last 2017). Artificial intelligence (AI) algorithms, for instance, "trained" on historical huge data and reflect the ideals and preconceptions of their creators as well as programmers (Montes and Goertzel 2019).

With the algorithms expanding globally, they amplify prejudices and perpetuate stereotypes against the most marginalized people. It becomes essential to include ethical and human values in technologies. In fact, the separation between technological and social advancement is no longer viable in a setting of widening wealth disparities among workers. New paradigms and stories must be developed immediately for inclusive prosperity (Last 2017). The term "neurodiversity" in psychology refers to the amalgamation of benefits and drawbacks brought on by a person's unique brain makeup. These variations include disorders, including attention deficit hyperactivity disorder (ADHD), autism spectrum disorder, dyslexia, and dyspraxia, among others. As autistic workers are presently in the spotlight for the majority of programs pertaining to employment that favor neurodiversity, the word "neurodiversity" is used in this research to more precisely refer to those individuals. Autism is a neurological psychiatric illness that lasts a lifetime and affects people differently in terms of perception and cognition. The neurodiversity paradigm takes into account the intangible distinctions between individual brains and intelligences, while the majority of literature based on diversity, incorporation, or inequality highlights overt characteristics like gender, age, or race.

This study offers a novel conceptual framework for examining the different connections between neurodiversity and the digital transition, which researchers view as a complementary pair. Despite their close connections, neurodiversity management and the digital transition have not yet been thoroughly researched.

3.11.1 The links between technology, organization, and skills

According to the skill-bias technological change theory (SBTC) (Acemoglu and Autor 2011), based upon the extent of complementarity among various inserts, the labor division between employees and machines, defined by their skill sets, is altered as a result of technological development. Since it disregards the organizational spectrum of the job, SBTC theory is deterministic. According to technological determinism, workstation organization and the distribution of work load between employees and technologies may be dictated by technology. In order to assert that developments, training, and institutional changes are necessary to create gains in terms of productivity with the involvement of new technologies, scientists are also taking note of the institutional aspect of technology-based progress (Camiña, Díaz-Chao, and Torrent-Sellens 2020). In other words, the implementation of technologies, expertise based on skill set, and supervisory design practices all work well together. In terms of microeconomics, when the reciprocal variables pertaining to a function escalate at the same time, the value of that particular function escalates higher than the summation of the changes that are brought about by the escalation in every variable when considered separately. Also, this strategy is very acceptable. Economic acceptance takes the place of technological determinism. The coordination of technical and organizational decisions is motivated by the pursuit of performance (Lambrecht and Tucker 2019). Reversing the loss to examine the influence of human acceptance in the technology framework is another method for highlighting biases in technologies. Instead of focusing on how technologies affect the workforce, let's examine how the makeup of the workforce affects technologies. Biases in the values and traits contained in technologies are created by technology designers. While the obvious gender imbalance in the IT sector is receiving a lot of attention, algorithmic biases related to impairments also raise additional concerns about social norms and stereotypes as well as job impediments. Initiatives to promote neurodiversity address the issue of preconceptions in the workplace while aiming for a more inclusive hiring process (Lambrecht and Tucker 2019).

3.11.2 Problem analysis

Researchers have devised two important and connected ideas, generally known as skill bias technology and organizational change (SBTC and SBOC) and productive interrelations, to explain the developing disparities at work associated with technological change.

These ideas suggest that a worker's abilities or degree of competence serve as the basis for comparison. Homogeneous workers possess heterogeneous talents or abilities. Obviously, additional factors distinguish workers when they are hired or promoted. On the job market and at work, stereotypes and discrimination against workers are prevalent. It is essential to take into account the workforce diversity, or the identity of the employees. The observational approach is applied by taking into account the particular traits, propensities, and stereotypes associated with autism. A field experiment demonstrated that appeals for considering jobs that said an employee had Asperger syndrome and autism obtained 26% lesser expressions of significance for a similar standard of skill (Ameri et al. 2018). The SBOC, SBTC, and fruitful reciprocal studies are used to frame the connections between digital revolution and the workforce's neurodiversity in ensuing subsections.

3.11.3 Neurodiversity at the workplace at different levels

Neurodiversity recognizes that each person's uniqueness as a human being is marked by the diversity of their perceptual and mental abilities. Understanding human variety is essential for comprehending neurodiversity. Consider the example of the intellectual quotient (IQ) Gaussian distribution, which indicates that 95% of the population has an IQ between

70 and 130. Individuals in the tails of the distribution differ from the mean or median because they deviate statistically from it. Each person is clearly defined by a variety of cognitive, emotional, and perceptual characteristics. But for the majority of people, the differences between two people pale in comparison to what they have in common. These differences are greater for autistic people, which beg the query of how these individuals specifically add up to the organization.

Both at the macro- and microeconomic levels, neurodiversity practices are acknowledged to have good social and inclusionary effects (Krzeminska et al. 2019). Apart from reputational advantages, transnational as well as major companies from various organizations have pioneered a change in how these individuals take charge of their staff and have created programs to utilize hitherto untapped autistic abilities. As a result, these organizations are actually acquiring a competitive benefit because of these creative projects (Austin and Pisano 2017). More businesses of all sizes and in a variety of industries are adopting neurodiversity hiring policies (Austin and Sonne 2014).

At the level of the individual, characteristics defining exceptional autistic employees involve their capacity for concentration, pattern recognition, ability to accomplish monotonous tasks, remarkable attention to detail, and participation. This may be the reason why many businesses that specialize in the placement of autistic individuals provide software-testing positions or coding jobs. According to a study, certain autistic people may be well suited for these positions since they have good rule-based system-building skills and a low mistake probability on activities requiring close attention to detail. The American Psychiatric Association (2013) noted that reoccurring motif hobbies or pursuits may also be a strength for autistic professionals who complete repetitive tasks in their field of interest. However, digitalization could have a negative effect on these jobs due to the automation of repetitive operations and the replacement of skilled jobs based on rules by artificial intelligence. This observation calls into question whether autistic workers' skills can be replaced by machinery. In fact, repetitive tasks and/or those requiring pattern recognition may be more susceptible to automation (substitution).

The establishment of an organization that accepts employees on the spectrum of autism is necessary for their employment and engagement at the organizational level, creating a workforce of neurodiverse employees (S.Markel & Elia, 2016). In a study, a variety of concerns and strategies are highlighted that have been used in various national, organizational, and institutional settings to promote the employment of autistic employees. According to (Austin and Pisano 2017), there are seven essential steps in the process of identifying neurodiversity in the worksite: the creation of non-stereotypical hiring practices; training of employees and employers; customization of the supervisory environment; specification of techniques for career management; guidance and counselling from experts having deep knowledge of neurodiversity in order to gain a clear understanding relating to the specifications of the employees; creations related to scaling strategies; and mainstreaming the program.

It is not a surprise that the IT sector receives the majority of attention in the articles pertaining to the effective handling of neurodiversity at the corporate level (Austin and Pisano 2017). In a field that moves quickly and where talents are continuously changing, the issue of a skill shortage is one that is frequently raised. Our study advances knowledge related to significance of gap in digital skills, the dynamism in IT job dismantling, and their possible connections to the adoption of neurodiversity initiatives.

The lack of digital skills is cited as a major motivator for neurodiversity initiatives in academia management research (Austin and Pisano 2017), although there is no clear evidence of a scarcity of digital and technical skills. According to Cappelli (2015), employers are primarily responsible for creating mismatches involving the ability gap between demand, supply, and shortage of skill sets, as a result of inadequate vocational guidance at the organization. This is an indication that the scarcity of digitally skilled workers is endogenous, resulting from corporate behavior, rather than exogenous to businesses. In fact, unfair hiring practices and discrimination restrict the pool of viable candidates and could lead to a scarcity of digital skills. As a result, encouraging neurodiversity at the corporate level is crucial for finding untapped sources of creative genius.

3.11.4 Methodology

This exploratory study seeks to determine the linkage between digital transition and neurodiversity management. The research was shaped by a phenomenological approach. A qualitative research method known as phenomenology describes participants' actual experiences in order to better comprehend their nature or significance. To accomplish the goals of the study, researchers employed a purposive sample strategy in accordance with the phenomenological approach. To find suitable interview candidates, researchers looked at two factors: leadership or competence in neurodiversity efforts, and solid IT industry knowledge (Walkowiak 2021).

In order to gather information with a suitable level of complexity and diversity, sixteen candidates were interviewed in 2018 and 2019. As per the participant expertise and contact intensity, this number in phenomenology typically ranges from 5 to 30 people. Sixteen participants offered the amount of saturation that was sought for, given the research's targeting of specialists, and the nascent approach of workplace initiatives in neurodiversity. Participants are listed in Table 3.1: *Participants*.

Three sets of questions served as the framework for the quasi-interviews. Identifying autistic workers' skills and how they relate to the digital transition was the goal of the first set of questions. Second series of queries centered on highlighting advantages as well as disadvantages pertaining to autistic employees related to innovation, artificial intelligence, and the shortage of

Table 3.1 Participants

P No.	Field	Gender	Country
1.	Financial	F	Canada
2.	Information Technology	M	Australia
3.	Social Enterprise	M	America
4.	Software	F	Canada
5.	Specially-abled Services	F	Canada
6.	Information Technology	M	Britain
7.	Information Technology	F	Australia
8.	Academic	M	Australia
9.	Charity Trust Institutes	M	Australia
10.	Accountancy	M	America
11.	Information Technology	M	Australia
12.	Software	F	Britain
13.	Social Enterprise	M	Canada
14.	Specially-abled Services	M	Canada
15.	Academic	F	Australia
16.	Information Technology	M	America

programmed skills. The final array of queries focused on the organizational architecture pertaining to neurodiversity's digital component.

An integrated method was implemented to code the content. It is less restricting than a start list method and less biased than an entirely induction-based method evolved from grounded theory. The broad code types were started, and further data-based sub-codes were created. The initial types of broad code covered the compatibility between autistic workers' abilities and programmed technologies, connections between neurodiversity-aware Human Resource practices as well as technology utility, and ultimately the dynamics belonging to hiring especially abled employees in conjunction with mechanization, AI, and program technologies (Figure 3.5: *Proposed Research Framework*). Researchers found various productivity factors that control the interactions between digital changeover and the employee's neurodiversity (Walkowiak 2021).

3.11.5 Result

While discussing a range of interests and abilities they saw in a neurodiverse workforce, participants also explicitly referred to noteworthy "performative" qualities that were connected with the analyzed output of an employee when he/she was carrying out an IT oriented task. Resilience, a different mode of innovative thinking as well as query resolving, which promotes creativity in the workplace and aids in the digital transformation, was one of these expressive capabilities.

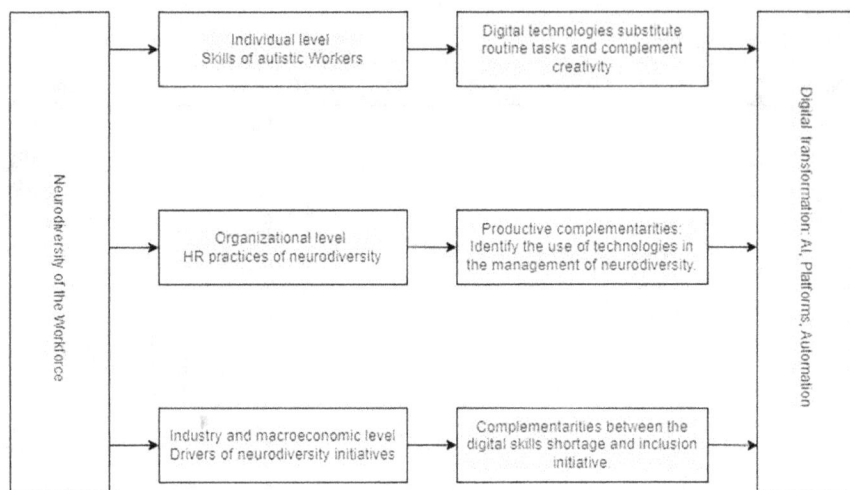

Figure 3.5 Proposed research framework.

In addition to technical qualifications, STEM, math, and computer science qualifications, which are connected to the programmed transition, all candidates cited a wide spectrum of capabilities and qualifications displayed by autistic employees. Additionally, they mentioned expertise in games, languages, the arts, history, literature, human resources, communication, budgeting, medical, and film. The diverse skill set displayed by autistic employees implies that they may be suitable for a range of occupations. Participants talked about a range of interests and abilities they saw in autistic employees, but they also methodically identified a few standout skills. Participants mentioned a variety of abilities, but they also noted that some of these skills—or how they are combined—might be very different from person to person. These abilities included emphasis to detail, a constant capacity for focus as well as discipline, firm involvement, the capability to recognize patterns, knowledge-grasping caliber, and innovative brain storming. Five participants felt that highlighting these particular skills possessed by autistic people was undesirable because it would help to reinforce negative preconceptions about autistic workers. These preconceptions, according to participant 16, are a result of the lack of leaders from the autistic community and social variations on squad in charge of neurodiversity projects. The creation of "ergonomic" environments that reduce the sensory challenges faced by autistic employees is a crucial component of neurodiversity projects' success. Workspace placement in locations with little stimuli promotes concentration. Additionally, relatively basic technology that enhances the sensory experience for autistic workers includes headsets and various light colors (Walkowiak 2021).

3.12 SCOPE AND CONCLUSION

The combined scope of artificial intelligence, neuroscience, and computational systems biology is significantly broad. The motivation of this unique difficulty is to create a drone view on areas and demanding situations in which the three fields overlap with their defining aims and where these fields might also benefit from a synergetic mutual exchange of ideas. The purpose in the back on this special issue is that a multidisciplinary technique in present-day artificial intelligence, neuroscience, and systems biology is vital and that progress in these fields requires a mess of perspectives and contributions from a huge spectrum of members. This special difficulty, therefore, objectives to create a center of gravity pulling together instructional researchers and industry practitioners from a selection of regions and backgrounds to percentage outcomes of modern-day studies and improvement and to talk about existing and rising theoretical and realistic troubles in synthetic intelligence, neuroscience, and structural biology, transporting them beyond the event horizon in their individual domain names (Berrar, Sato, and Schuster 2010).

Bioscience and neuroscience are both devoted to growing our information of the human frame and its different aspects, with a mutual interest in the use of expertise to enhance human existence. These fields significantly diverge, however, due to their focus on observation. Questions that arise in biomedicine (and finally bioethics) are commonly constrained to the area of clinical research and clinical motion. While neuroscience has its analogues to those questions, its scope extends nicely beyond the confines of clinical exercise: the mind's precise function in behavior engenders questions in a spectrum of fields with significant effects on society, such as neuro-politics, neuro-economics, and neuro-regulation (McCoy et al. 2020).

REFERENCES

Abbott, L. F. 2008. "Theoretical Neuroscience Rising." *Neuron* 60 (3): 489–495. 10.1016/j.neuron.2008.10.019.

Abdel-Nasser Sharkawy. 2020. "Principle of Neural Network and Its Main Types: Review." *Journal of Advances in Applied & Computational Mathematics* 7: 8–19. 10.15377/2409-5761.2020.07.2.

Abdelhameed, Magdy M, and Farid A Tolbah. 2002. "Design and Implementation of a Flexible Manufacturing Control System Using Neural Network." *International Journal of Flexible Manufacturing Systems* 14 (3): 263–279. 10.1023/A:1015883728142.

Acemoglu, Daron, and David Autor. 2011. *Skills, Tasks and Technologies: Implications for Employment and Earnings. Handbook of Labor Economics.* Vol. 4. Elsevier B.V. 10.1016/S0169-7218(11)02410-5.

Ameri, Mason, Lisa Schur, Meera Adya, F. Scott Bentley, Patrick McKay, and Douglas Kruse. 2018. "The Disability Employment Puzzle: A Field Experiment on Employer Hiring Behavior." *ILR Review* 71 (2): 329–364. 10.1177/0019793917717474.

Austin, Robert D., and Gary P. Pisano. 2017. "Neurodiversity as a Competitive Advantage." *Harvard Business Review* 2017 (May-June): 9.

Austin, Robert D., and Thorkil Sonne. 2014. "The Dandelion Principle: Redesigning Work for the Innovation Economy." *MIT Sloan Management Review* 55 (4): 67–72.

Barak, Omri. 2017. "Recurrent Neural Networks as Versatile Tools of Neuroscience Research." *Current Opinion in Neurobiology* 46: 1–6. 10.1016/j.conb.2017.06.003.

Barak, Omri, David Sussillo, Ranulfo Romo, Misha Tsodyks, and L. F. Abbott. 2013. "From Fixed Points to Chaos: Three Models of Delayed Discrimination." *Progress in Neurobiology* 103: 214–222. 10.1016/j.pneurobio.2013.02.002.

Berns, Gregory S, Jonathan D Cohen, and Mark A Mintun. n.d. "Absence of Awareness."

Berrar, Daniel, Naoyuki Sato, and Alfons Schuster. 2010. "Artificial Intelligence in Neuroscience and Systems Biology: Lessons Learnt, Open Problems, and the Road Ahead." *Advances in Artificial Intelligence* 2010: 1–2. 10.1155/2010/578309.

Camiña, Ester, Ángel Díaz-Chao, and Joan Torrent-Sellens. 2020. "Automation Technologies: Long-Term Effects for Spanish Industrial Firms." *Technological Forecasting and Social Change* 151 (November 2019): 119828. 10.1016/j.techfore.2019.119828.

Caspi, Avshalom, Joseph McCray, Terrie E. Moffitt, Jonathan Mill, Judy Martin, Ian W. Craig, Alan Taylor, and Richie Poulton. 2002. "Role of Genotype in the Cycle of Violence in Maltreated Children." *Science* 297 (5582): 851–854. 10.1126/science.1072290.

Cox, David Daniel, and Thomas Dean. 2014a. "Neural Networks and Neuroscience-Inspired Computer Vision." *Current Biology* 24 (18): R921–R929. 10.1016/j.cub.2014.08.026.

Cox, David Daniel, and Thomas Dean. 2014b. "Neural Networks and Neuroscience-Inspired Computer Vision." *Current Biology* 24 (18): R921–R929. 10.1016/j.cub.2014.08.026.

Dwyer, Patrick. 2022. "The Neurodiversity Approach(Es): What Are They and What Do They Mean for Researchers?" *Human Development* 66 (2): 73–92. 10.1159/000523723.

Enel, Pierre, Emmanuel Procyk, René Quilodran, and Peter Ford Dominey. 2016. "Reservoir Computing Properties of Neural Dynamics in Prefrontal Cortex." *PLoS Computational Biology* 12 (6): 1–35. 10.1371/journal.pcbi.1004967.

"Fundamentals of Machine Learning and Softcomputing." 2006. *Neural Networks in a Softcomputing Framework*, no. Vc: 27–56. 10.1007/1-84628-303-5_2.

Gao, Peiran, and Surya Ganguli. 2015. "On Simplicity and Complexity in the Brave New World of Large-Scale Neuroscience." *Current Opinion in Neurobiology* 32: 148–155. 10.1016/j.conb.2015.04.003.

Geake, John, and Paul Cooper. 2003. "Cognitive Neuroscience: Implications for Education?" *Westminster Studies in Education* 26 (1): 7–20. 10.1080/0140672030260102.

Graben, Peter beim, and James Wright. 2011. "From McCulloch-Pitts Neurons Toward Biology." *Bulletin of Mathematical Biology* 73 (2): 261–265. 10.1007/ s11538-011-9629-5.

Hopfield, J. J. 1982. "Neural Networks and Physical Systems with Emergent Collective Computational Abilities." *Proceedings of the National Academy of Sciences of the United States of America* 79 (8): 2554–2558. 10.1073/ pnas.79.8.2554.

Houting, Jacquiline den. 2019. "Neurodiversity: An Insider's Perspective." *Autism* 23 (2): 271–273. 10.1177/1362361318820762.

Hyvärinen, Aapo. 2010. "Statistical Models of Natural Images and Cortical Visual Representation." *Topics in Cognitive Science* 2 (2): 251–264. 10.1111/j.1756-8765.2009.01057.x.

Jaeger, Herbert, and Harald Haas. 2004. "Harnessing Nonlinearity: Predicting Chaotic Systems and Saving Energy in Wireless Communication." *Science* 304 (5667): 78–80. 10.1126/science.1091277.

Kapp, Steven K. 2020. *Conclusion - Autistic Community and the Neurodiversity Movement: Stories from the Frontline. Autistic Community and the Neuro-diversity Movement: Stories from the Frontline.* 10.1007/978-981-13-8437-0_22.

Katiyar, Kalpana. 2022. "AI-Based Predictive Analytics for Patients' Psychological Disorder." In *Predictive Analytics of Psychological Disorders in Healthcare: Data Analytics on Psychological Disorders*, 37–53. Springer.

Katiyar, Kalpana, Pooja Kumari, and Aditya Srivastava. 2022. "Interpretation of Biosignals and Application in Healthcare." In *Information and Communication Technology (ICT) Frameworks in Telehealth*, 209–229. Springer.

Katiyar, Sarthak, and Kalpana Katiyar. 2021. "Recent Trends Towards Cognitive Science: From Robots to Humanoids." In *Cognitive Computing for Human-Robot Interaction*, 19–49. Elsevier.

Krzeminska, Anna, Robert D. Austin, Susanne M. Bruyère, and Darren Hedley. 2019. "The Advantages and Challenges of Neurodiversity Employment in Organizations." *Journal of Management and Organization* 25 (4): 453–463. 10.1017/jmo.2019.58.

Lambrecht, Anja, and Catherine Tucker. 2019. "Algorithmic Bias? An Empirical Study of Apparent Gender-Based Discrimination in the Display of Stem Career Ads." *Management Science* 65 (7): 2966–2981. 10.1287/mnsc.2018.3093.

Last, Cadell. 2017. "Global Commons in the Global Brain." *Technological Forecasting and Social Change* 114 (2016): 48–64. 10.1016/j.techfore.2016.06.013.

Machens, Christian K., Ranulfo Romo, and Carlos D. Brody. 2005. "Flexible Control of Mutual Inhibition: A Neural Model of Two-Interval Discrimination." *Science* 307 (5712): 1121–1124. 10.1126/science.1104171.

Mante, Valerio, David Sussillo, Krishna V. Shenoy, and William T. Newsome. 2013. "Context-Dependent Computation by Recurrent Dynamics in Prefrontal Cortex." *Nature* 503 (7474): 78–84. 10.1038/nature12742.

Material, Supplementary. 2002. "Caspi_2002_MAOA_AggressionMsat_SIMethods," 1–7. papers2://publication/uuid/D97CED7A-D777-4E60-A883-A49E4CE1B625.

McCalpin, John D. 1995. "Memory Bandwidth and Machine Balance in Current High Performance Computers." *IEEE Computer Society Technical Committee on Computer Architecture (TCCA) Newsletter*, no. May: 19–25.

McCoy, Liam G., Connor Brenna, Felipe Morgado, Stacy Chen, and Sunit Das. 2020. "Neuroethics, Neuroscience, and the Project of Human Self-Understanding." *AJOB Neuroscience* 11 (3): 207–209. 10.1080/21507740.2020.1778127.

Montes, Gabriel Axel, and Ben Goertzel. 2019. "Distributed, Decentralized, and Democratized Artificial Intelligence." *Technological Forecasting and Social Change* 141 (February): 354–358. 10.1016/j.techfore.2018.11.010.

Ogden, Thomas E., and Robert F. Miller. 1966. "Studies of the Optic Nerve of the Rhesus Monkey: Nerve Fiber Spectrum and Physiological Properties." *Vision Research* 6 (10). 10.1016/0042-6989(66)90108-8.

Parra, Miguel, George Hruby, and George Hruby. n.d. "Neuroscience and Education Related Papers."

Rivkind, Alexander, and Omri Barak. 2017. "Local Dynamics in Trained Recurrent Neural Networks." *Physical Review Letters* 118 (25): 1–5. 10.1103/PhysRevLett. 118.258101.

S. Markel, Karen, and Brittany Elia. 2016. "How Human Resource Management Can Best Support Employees with Autism: Future Directions for Research and Practice." *Journal of Business and Management* 22 (1): 71–86.

Sharkawy, Abdel-Nasser, Panagiotis N Koustoumpardis, and Nikos Aspragathos. 2020. "A Neural Network-Based Approach for Variable Admittance Control in Human-Robot Cooperation: Online Adjustment of the Virtual Inertia." *Intelligent Service Robotics* 13 (4): 495–519. 10.1007/s11370-020-00337-4.

Srivastava, Aditya, and Shashank Jha. 2022. "Data-Driven Machine Learning: A New Approach to Process and Utilize Biomedical Data." In *Predictive Modeling in Biomedical Data Mining and Analysis*, 225–252. Elsevier.

Srivastava, Aditya, Aparna Seth, and Kalpna Katiyar. 2021. "Microrobots and Nanorobots in the Refinement of Modern Healthcare Practices." In *Robotic Technologies in Biomedical and Healthcare Engineering*, 13–37. CRC Press.

Sussillo, David. 2014. "Neural Circuits as Computational Dynamical Systems." *Current Opinion in Neurobiology* 25: 156–163. 10.1016/j.conb.2014.01.008.

Szegedy, Christian, Wojciech Zaremba, Ilya Sutskever, Joan Bruna, Dumitru Erhan, Ian Goodfellow, and Rob Fergus. 2014. "Intriguing Properties of Neural Networks." *2nd International Conference on Learning Representations, ICLR 2014 - Conference Track Proceedings*, 1–10.

Thibault, Jules, and Bernard P.A. Grandjean. 1991. "A Neural Network Methodology for Heat Transfer Data Analysis." *International Journal of Heat and Mass Transfer* 34 (8): 2063–2070. 10.1016/0017-9310(91)90217-3.

Walkowiak, Emmanuelle. 2021. "Neurodiversity of the Workforce and Digital Transformation: The Case of Inclusion of Autistic Workers at the Workplace." *Technological Forecasting and Social Change* 168 (April): 120739. 10.1016/j.techfore.2021.120739.

Wang, Hongming, Ryszard Czerminski, and Andrew C. Jamieson. 2021. "Neural Networks and Deep Learning." *The Machine Age of Customer Insight*, 91–101. 10.1108/978-1-83909-694-520211010.

White, B. W., and Frank Rosenblatt. 1963. "Principles of Neurodynamics: Perceptrons and the Theory of Brain Mechanisms." *The American Journal of Psychology* 76 (4): 705. 10.2307/1419730.

Chapter 4

Brain waves, neuroimaging (fMRI, EEG, MEG, PET, NIR)

Surbhi Kumari and Amit Kumar Dutta

Amity Institute of Biotechnology, Amity University Jharkhand Ranchi, India

4.1 INTRODUCTION

Neuroscience is the integrative science dealing with the study of the nervous system (both central and peripheral nervous system) as well as its functions and underlying diseases. The brain is the vital organ of the central nervous system controlling all the other activities in the body. Computational neuroscience is the study of the mechanism of brain functioning by various tools and techniques using computer science (Sejnowski et al., 1988).

Our complex brain consists of fundamental units called neurons. The functional neuron in the brain transmits the electrical signals. There are on an average 90 billion neurons present in the human brain (Goriely et al., 2015). The electrochemical signals are carried by these neurons and result in passing of the iconic currents via the synapse. The coordination of the electrical activities of the neurons results in the repeated rhythmic alterations all over the regions of brain and are termed as brain waves (Buskila, 2019). (Figure 4.1)

The five brain waves that have been recognized to date are: gamma (g), beta (b), alpha (a), theta (θ), and delta (d) waves. All these waves possess different frequencies, for e.g., gamma waves have the highest frequency range (<30 Hz), whereas delta waves have the lowest frequency range (0–4 Hz). They all are responsible for different states of mind in the human brain (Abhang et al., 2016). These brain waves are measured by a technique known as electroencephalography, or EEG, that uses electrodes placed on the scalp for recording the waves by non-invasive approach (Teplan, 2002). The brain structure and function can be easily analyzed by studying the images of the brain by the process of neuroimaging. It has been broadly classified into structural as well as functional neuroimaging techniques. The functional techniques like fMRI, PET, MEG, and NIR are covered in the chapter (Noggle & Davis, 2021). Numerous methods of non-invasive neuroimaging have fostered an enormous impact in upgrading our knowledge about the brain functioning and neurological disorders and provided a novel insight into the treatment of these diseases (Supek & Aine, 2016).

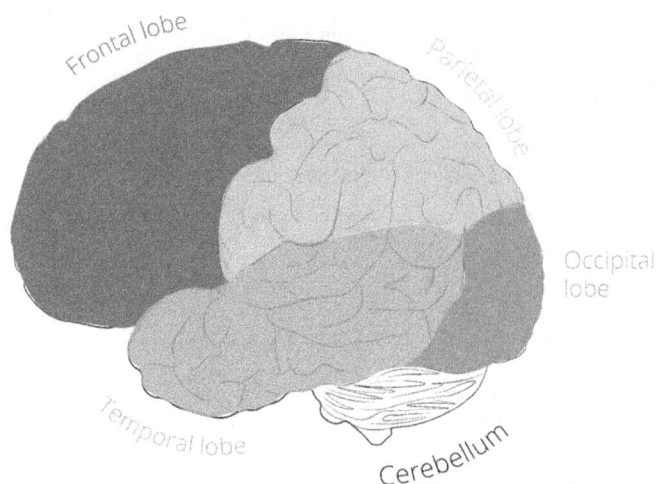

Figure 4.1 The different lobes of the brain viz. frontal, parietal, occipital and temporal lobe, that possess diverse functions like controlling emotions, processing sensory data and visual information, and conducting auditory functions, respectively.

(https://imotions.com/wp-content/uploads/2022/10/brain-lobes-iMotions.png).

4.2 BRAIN WAVES

The synchronized wave patterns generated by the flow of electric current through the neurons are called brain waves (Ismail et al., 2016). The voltage variations in the neurons occur due to the ionic flow within the brain. These waves come about naturally while in mobile as well as in a relaxed phase. They are determined using an EEG. These electrical activities of the brain are recorded using the placement of metal electrodes on the scalp to detect the distinct frequencies of the brain waves viz. gamma, beta, alpha, theta, and delta (Jeong et al., 2011) (Figure 4.2).

a. *Gamma (γ) Brain waves* – The gamma brain waves are the fastest waves, having the largest frequency range above 30 Hz (i.e., more than 30 cycles per second). They govern while the brain is involved in analytical problem solving (Jeong et al., 2011), deep learning, as well as creative processing of language (Ismail et al., 2016). These waves have emerged from the thalamic region of the brain. They are initiated to adjust the activity and systems of the neurons. People with some injury in the thalamus lack consciousness along with cognition, and as a result may slip into a coma (Desai et al., 2015). There should be balance in gamma waves for the right perception, cognitive focus, inspired learning, and processing of information. Lowered levels result in impairment in learning, depression, as well as ADHD. On the other hand, increased

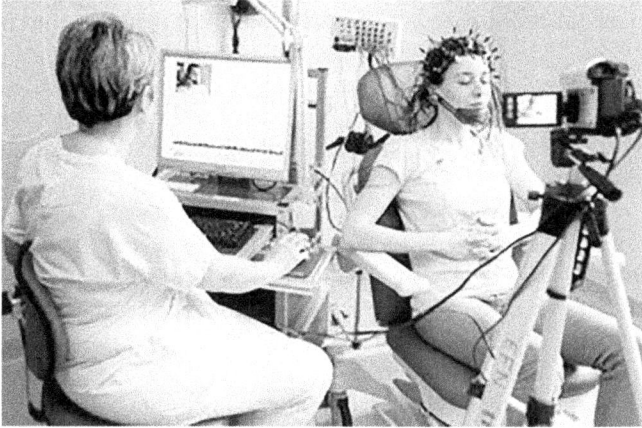

Figure 4.2 A non-invasive process of electroencephalography (EEG) for the analysis and computation of brain waves.

(https://assets.nhs.uk/nhsuk-cms/images/E5RF5X.width-1534.jpg).

levels lead to stress, hyperactivity, and anxiousness (Dudeja, 2017). In research, it was found that people practicing meditation had increased gamma activation frequencies. The brain corresponds to the occurring stream of consciousness and attention, among other cognitive functions, which have all been related to gamma waves (Braboszcz et al., 2017). They are regulated via interior processes like attention as well as working memory. Its level elevates with the cognitive and noncognitive events along with the sensory drive. In disorders associated with the central nervous system like Alzheimer's, epilepsy and Parkinson's disease, the gamma activities have been seen as irregular (Jia & Kohn, 2011).

b. *Beta (β) Brain Waves* – Beta waves lie between the frequency range 14–30 Hz (i.e., 14–30 cycles per second). These waves act throughout the waking consciousness (Jeong et al., 2011). They are highly active during problem solving like logical or analytical reasoning (Koudelková et al., 2018), initiating new ideas, and generating solutions. They promote elevated concentration levels and alertness of our brain towards tasks (Ismail et al., 2016). Optimal amount of beta waves leads to better focus, memory and ability to solve a problem. When the frequency is low, it leads to a state of depression, deficit cognition, daydreaming and disorders like attention-deficit hyperactivity disorder, or ADHD. On the other hand, too much of it causes increased stress, anxiety and even hyperactivity with a distracted mind (Dudeja, 2017). These waves are present all over the motor region of the cortex during the muscular isotonic contractions. The triggering of the beta waves corresponds to better performance in academics due to improved cognition as well as concentration (Desai et al., 2015).

The alert mind is the result of beta waves that in turn result in effectual functioning (Ashtaputre-Sisode, 2016).

c. *Alpha (α) Brain Waves* – Alpha waves are also called Berger's wave as they were manifested by Hans Berger in the 1930s (Stinson & Arthur, 2013). The conscious thinking in addition to the subconscious mind are interconnected by the alpha waves. Their frequency lies between 8–13 Hz (i.e., 8–13 cycles per second). They assist the feeling of relaxation and calm down the body (Koudelková et al., 2018). They are initiated inside the cortex, near the thalamus and the occipital lobe (Desai et al., 2015). The alpha waves govern the state of light meditation or daydreaming (Jeong et al., 2011), when the mind is relaxed, and thoughts are passing over. The requirement of a stabilized amount of alpha waves is necessary because if the quantity is less it results to insomnia, stress, and obsessive compulsive disorder. If the quantity is more it leads to less concentration and complete relaxed mind (Dudeja, 2017). These waves intensify the learning process and the physical as well as mental health of a person. It also aids in the therapy of depression, anxiety, and sleep dysfunctions by introducing the deep relaxation approach along with mindfulness (Stinson & Arthur, 2013). In older adults these waves ameliorate the identification of words and enable precise memory performance. The person behaves in a calm and composed manner (Desai et al., 2015). A study revealed that the person suffering from anxiety disorder had an increased frequency of alpha waves on the frontal lobe (both sides) as well as on the parietal lobe (front side). The disorders of anxiety are basically related to the dysfunction of the forebrain (Cho et al., 2011) (Figure 4.3).

d. *Theta (θ) Brain Waves* – The frequency of the Theta waves ranges between 4–8 Hz (i.e., 4 to 8 cycles per second). They are dominant throughout normal sleeping hours, while meditating (Jeong et al., 2011) and during states of fatigue (Ashtaputre-Sisode, 2016). Walter and Dovey first described the presence of theta waves in 1944 in cases associated with tumors in the sub-cortex (Schacter, 1977). They result in profound meditation, relaxation, and enhanced memory in individuals. During light sleep or conscious dreaming in daytime, they get activated and correlate with relieving stress (Ismail et al., 2016). Theta rhythm is the other name given to these waves, which eventuate as a repeated function turns to be self-governed. These waves originate in the region of the cortex as well as hippocampus. It was suggested in a study that theta waves are associated with constructing memories via action within the hippocampus (Desai et al., 2015). While we are dreaming or in a state of intuition, unconsciousness, or imagination, these waves are governing our mind. Their excessive number results in disorders like ADHD, distracted mind, impulsiveness, depression, and hyperactivity disorders. Decreased amount of these waves causes stress, anxiousness, and deficit emotional balance. Adequate balance in waves

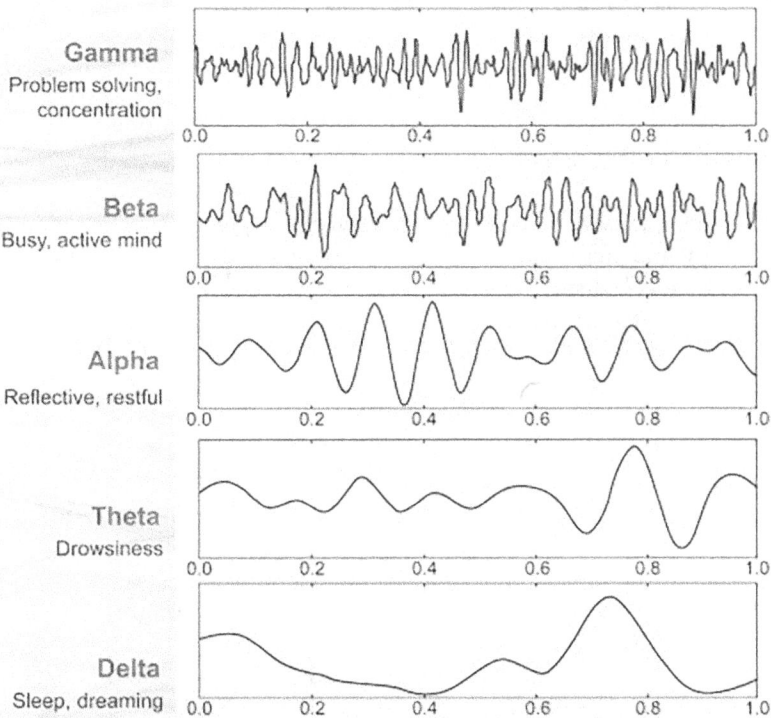

Figure 4.3 The different types of brain waves associated with their dominant states. **(https://ars.els-cdn.com/content/image/3-s2.0-B9780128044902000026-f02-01-9780128044902.jpg).**

leads to proper flow of thoughts, emotional stability, being intuitive and relaxed (Dudeja, 2017). During hypnosis state, theta waves are active, and our brain is filled with thoughts of our inner self. In some research, it has been found that these waves are associated with psychological occurrences in humans (Schacter, 1977).

e. *Delta (δ) Brain Waves* – The slowest brain waves are the delta waves, with frequency ranging between 0–4 Hz (i.e., less than 4 cycles per second). During deep sleep or profound meditation, these waves are dominant as this deep sleep state is an essential and vital process that aids in healing and the process of regeneration. Excessive number of delta waves leads to brain trauma, disoriented mind, cognition disability and disorders like ADHD. On the contrary, fewer delta waves causes problems in sleep cycles, thinking, learning, and inability of the body to regenerate as well as revitalize itself (Dudeja, 2017). The Human Growth Hormone is released along with the delta waves, which aids in the healing process. If it is generated in a state of waking consciousness, it gives the chance to approach the subconscious state (Ismail et al., 2016). People with impairment associated with learning processes or brain injuries have unusual activity of the delta waves (Koudelková et al., 2018).

4.3 NEUROIMAGING

Neuroimaging is a non-invasive approach to get in touch with different structures and activities of our brain (Gui et al., 2010). It has been an emerging and indispensable procedure for recognizing the brain dynamics on a structural level (Bandettini, 2009). The functional dynamics of the brain can be encapsulated by neuroimaging by capturing the disparate timespans of the brain activities (Kringelbach & Deco, 2020). The different functional neuroimaging techniques, namely PET (positron emission tomography), MEG (magnetoencephalography), fMRI (functional magnetic resonance imaging), NIR (near infrared), provide remarkable probability in neurological disease detection and examination (Drevets, 2000). The fMRI is a non-invasive process that determines accurate brain activities with elevated geometric resolution. The PET is the significant technique for depicting the overall physiological features of the brain (Savoy, 2001) (Figure 4.4).

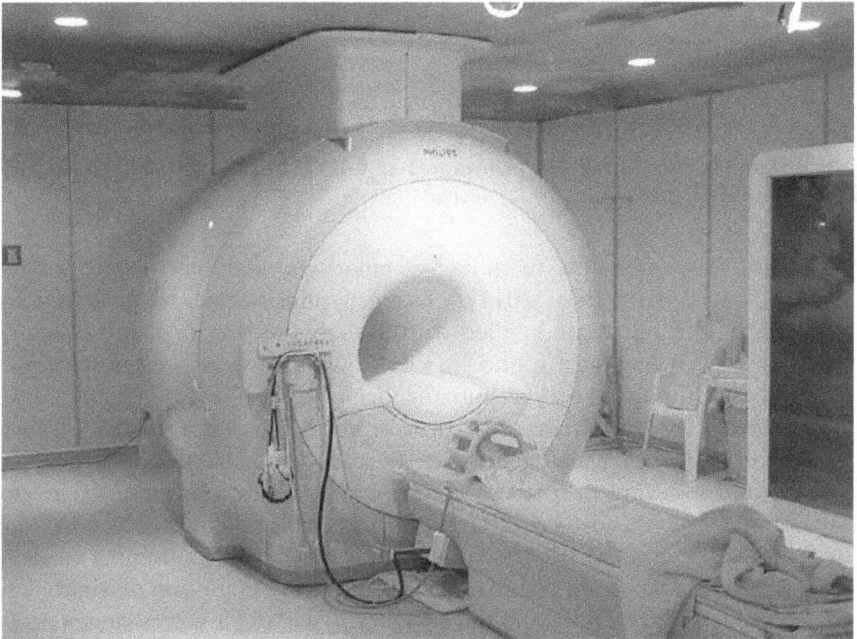

Figure 4.4 A magnetic resonance imagining (MRI) machine.

(https://ccnmtl.columbia.edu/projects/neuroethics/module1/foundationtext/ index.html).

a. *Functional Magnetic Resonance Imaging (fMRI)* – The fMRI is a non-invasive neuroimaging technique that has evolved to reveal the time-dependent variations of the metabolisms occurring in the brain (Glover, 2011). It has become a governing technique due to its least exposure to radiation, less invasive nature, and greater availability (Gui et al., 2010). From its establishment in 1992, it has enhanced cognitive studies for the healthy as well as the defective brain. It uses the MRI to represent the vital changes in the tissues underlying the brain generated by the variations in the neurological metabolisms (Chen & Glover, 2015). The fMRI is constructed on the principle of MRI that uses magnetic fields with their gradients and the radio waves for producing the images. It is basically associated with the learning of the cognitive behavior of the brain, diagnosis of certain neurological diseases, keeping track of different therapies, as well as depicting the efficacy of certain drugs (Figure 4.5).

The neural activities of the brain result in various metabolic actions such as rise in oxygen supply along with elongated blood flow, and these activities are detected using various techniques namely BOLD, perfusion, and contrast fMRI. Among these the BOLD (blood oxygenation level dependent) fMRI method is widely used (Gui et al., 2010). It maps the neurological

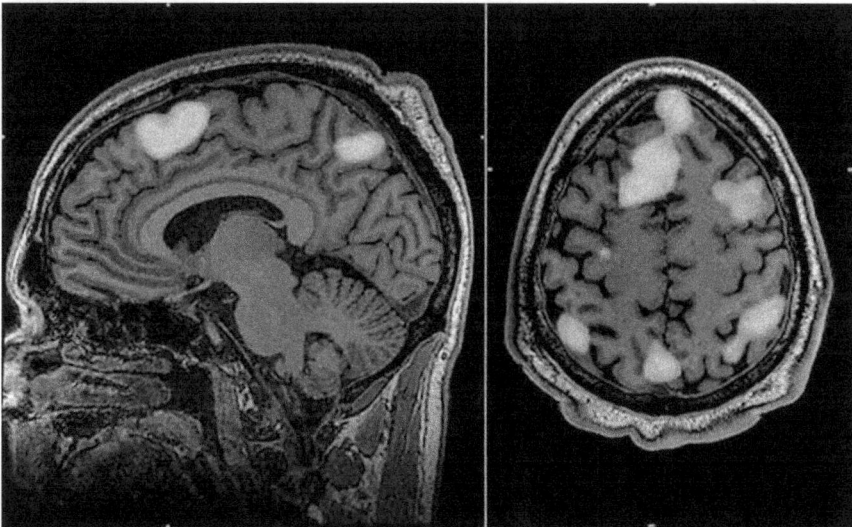

Figure 4.5 A fMRI scan of human brain depicting certain regions in a person suffering from severe traumatic brain injuries.

(https://imotions.com/blog/learning/research-fundamentals/eeg-vs-mri-vs-fmri-differences/ https://imotions.com/wp-content/uploads/2022/10/fMRI-explained.jpg).

activities by determining the variation in blood flow in the brain and determines the alteration in corresponding signal strength related to the various cognitive conditions amidst the imaging process (Matthews & Jezzard, 2004). There lies a limitation for the same as the detection of brain activities is not carried directly as it determines the variation in blood oxygenation level instead of directly measuring the neuron activity (Chen et al., 2020).

b. *Electroencephalography (EEG)* – The discovery of the brain possessing electrical currents was carried out by Richard Caton in 1875, whereas discovering these currents can be measured at the scalp surface was demonstrated by Hans Berger, who was a German neurologist, in 1924 (Teplan, 2002). EEG is a neuroimaging technique involving a non-invasive method to determine the electrical signals of the brain by making use of a cap containing metal electrodes and sensors that are laid on different positions on the scalp. EEG was first employed by Hans Berger in 1929, and in 1957 the use of electrodes was discovered in recording the activities of the brain by Gray Walter (Abhang et al., 2016). During the activation of neurons, the flow of current is generated that is measured by an EEG. The current perforates through various layers of the skin into the skull and then to the neurons as a result recording their electrical activities (Teplan, 2002). It records the brain waves of different frequencies in accordance with the variations of mental state (Abhang et al., 2016). Different types of electrodes are employed such as the gel-based electrodes, water-based electrodes, and dry electrodes (Figure 4.6).

They are positioned on a specific location. According to the 10–20 system, they are laid down at 10% and 20% from the left, right, nasion and inion

Figure 4.6 Different metal electrodes incorporated into the cap that is placed on the scalp to record the brain oscillations with contrasting frequencies.

(https://info.tmsi.com/blog/types-of-eeg-electrodes).

points, although to attain a spatial resolution, an additional number of electrodes are placed in this system (Müller-Putz, 2020). These metal electrodes scrutinize the signals at the surface of head, then they are amplified and are converted to digital pattern that is recorded on a computer (Teplan, 2002), and the voltage potentials generated by the flowing current in the neurons are recorded by the electrodes (Biasiucci et al., 2019). The EEG readings are provided to gain comprehensive information about the cognitive aspects, mental state such as stress, relaxation, etc. (Jebelli et al., 2018). In clinical diagnosis and treatment of neural rehabilitation it has its diverse range (Biasiucci et al., 2019). This neuroimaging method can detect the disorders of the brain such as epilepsy, tumors, brain strokes, injury, and dysfunction (Abhang et al., 2016).

c. *Magnetencephalography (MEG)* – MEG is a non-invasive functional neuroimaging technique that includes measuring the magnetic field produced by the electrical currents flowing through the neurons (Singh, 2014). The activities of neurons as recorded by the MEG technique are the result of both excited and inhibited postsynaptic potential of the dendrites. This flow of current produces a magnetic field projecting in a radial direction (Proudfoot et al., 2014). According to the right-hand rule in Ampere's law, electric current is related to the magnetic field that is perpendicular to the direction of current. The activated neurons produce electric current, which generates magnetic fields that get recorded with the help of MEG (Singh, 2014) (Figure 4.7).

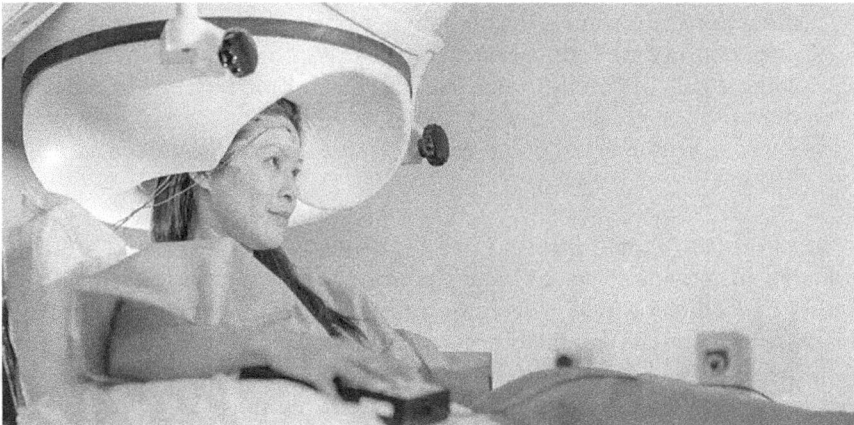

Figure 4.7 Magnetoencephalography neuroimaging process determining the magnetic field brought about by the electronic activity of the neurons in the brain.

(https://www.york.ac.uk/psychology/research/york-neuroimaging-centre/research/magnetoencephalography/).

Currently, MEG is the sole non-invasive imaging with higher-resolution technique that is independent from vascular responses. Furthermore, it is the only functional neuroimaging providing both elevated spatial as well as temporal resolution (Wilson et al., 2016). The MEG readings are capable of providing an insight regarding the processes and the functional framework of the brain. The magnetic fields determined by the MEG imaging are very minute; therefore, extremely sensitive sensors and detectors like SQUIDs or superconducting quantum interference devices are employed. Current MEG set-up comprises 100–300 sensors that envelope the head (Supek & Aine, 2016). The SQUIDs allow detection of very small magnetic field gradients with unit femtotesla or fT that are produced at the surface of the skull with the help of currents flowing in the cortex (Savoy, 2001). In the detection and study of several neurological disorders, MEG imaging has wide applications (Wilson et al., 2016). In patients with epilepsy, MEG aids in the identification of eloquent brain areas that helps in determining the affected region during surgery (Singh, 2014).

d. *Positron Emission Tomography (PET)* – The PET is a functional neuroimaging tool that aids in visualizing the physical as well as atomic activities in a specific region of the brain or different body parts (Raichle, 1983). This non-invasive technique is helpful in ameliorating knowledge about the pathophysiological aspects of a disease, diagnosis, and treatment results (Zimmer, 2009). It assists in the diagnosis of neurological disorders and underlying injuries of brain parts. The radioactive decay induces a huge amount of radiation, which is employed in PET (Raichle, 1983). This technique uses the radiotracers (radioactive particles) that cohere with the target molecule with greater bond strength and selectivity. These radiotracer molecules consist of an atom having immoderate quantity of nuclear energy generated because of its contrasting neutron count when compared to its stable state (Walker & Bilgel, 2021) (Figure 4.8).

Firstly, the radionuclides that emit positrons are generated and get included in molecules to give the radiotracers. These radionuclides produce PET radioisotopes. The radiotracers are administered in patients, which reaches to the intended organs, and positrons move and get united with the electrons, producing the photons that get captured by the camera and recorded via detectors. The images with temporal and spatial resolutions are generated and examined for diagnosis as well as treatment of diseases (Zimmer, 2009). Assessment of patients suffering from brain stroke is really complicated as a distinct region of the brain examined to check for permanent or partial damage caused, but with PET imaging it has been quite simple to picture a distinguished area for suitable therapy to be conducted. The same is true with a person with epilepsy, where this

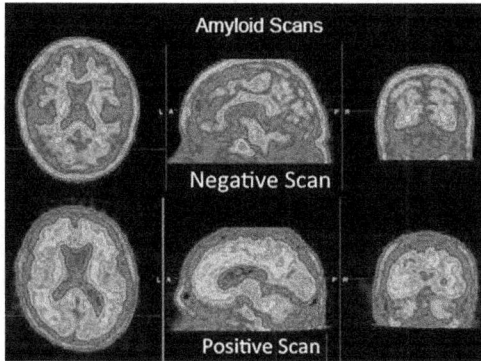

Figure 4.8 An amyloid PET scan (both positive and negative) depicting deposition of proteins such as tau and amyloid that forms plaques in the brain of a person suffering from Alzheimer's disease.

(https://radiology.ucsf.edu/patient-care/services/specialty-imaging/alzheimer).

technique is widely used to study the symptoms of seizures associated with epilepsy by detecting the affected functioning area of the brain through implementation of PET (Raichle, 1983).

Alzheimer's disease is a continuous degrading neurological ailment accompanied by weakened memory, inability to think, and inability to fulfil daily activities. For effective diagnosis of this disease, any of the two amyloid or FDG PET scannings are acceptable as biomarkers. In amyloid PET imaging the plagues causing damage to the nerve cells are visualized in the patient (Rice & Bisdas, 2017). PET neuroimaging also aids in scrutinizing the synthesis of proteins, process of Krebs cycle, glucose metabolism as well as DNA replications. The neurological activities of the brain and its hemodynamic variations are demonstrated using PET for e.g., the volume and flow of blood along with the use of oxygen (Savoy, 2001).

e. *Near-Infrared Spectroscopy (NIRS)* – NIRS is a non-invasive, functional neuroimaging procedure that includes close supervision of the alterations in the blood volume along with blood oxygenation associated with brain functioning. It makes use of certain light wavelengths that are exposed to the scalp surface to record the fluctuations in respective ratios of both oxygenated as well as deoxygenated haemoglobin throughout complete brain activity (Izzetoglu et al., 2005) (Figure 4.9).

However, the functional activities of the brain or any tissue can dominate its optical characteristics. In the same way, the human brain retaliates to the external stimuli and experiences certain physiological variations such as fluctuations in the level of blood along with the electrical activities, resulting in changes in optical characters, too. The oxygenated haemoglobin, as well as the deoxyhaemoglobin (oxy-Hb and deoxy-Hb)

Figure 4.9 NIRS imaging with head cap containing sensors, light detectors, and light source.

(https://www.pnas.org/doi/10.1073/pnas.2208729119).

acting as chromophores, comprises optical characters in both the near-infrared range and the visible range. The difference in concentration of both is determined using Beer-Lambert's Law (Irani et al., 2007). The deoxy-Hb chromophore absorbs below 790 nm, whereas the oxy-Hb absorbs above 790 nm. The NIRS signals respond to the variations in the neural activities that are entwined with the hemodynamic feedback functions (Chen et al., 2020).

Several NIRS studies have built strong insight in detection of neurological disorders in children as well as in adults like ADHD, Alzheimer's disease, epilepsy, migraine, etc. (Ehlis et al., 2014). Variations such as decrease in oxygenated Hb and total Hb in the frontal lobe were found in Alzheimer's patients. This suggests that there was gradual decrease in the oxygen supply in the degenerated regions of the brain, resulting in poor cognition in them. In recent studies, it has been observed that in epileptic patients the concentration of oxy-Hb and the blood volume increases profoundly during seizures (Irani et al., 2007).

4.4 CONCLUSION

The brain oscillations or brain waves are synchronized, and continuous patterns are generated due to the electrical activities of the neurons in the brain. They can be recorded easily via the process of EEG, which is a non-invasive technique. It includes several sensors and electrodes that are placed directly onto the scalp for recording the neuronal activities and mapping them on a computer system. These waves are categorized into five major types based on their frequencies and characteristics viz. gamma, beta, alpha, theta, and delta. Functional neuroimaging is a speedily evolving field in neuroscience. Diverse approaches and advancements have been made so far to the prevailing techniques. A few recent examples comprise of the establishment of fMRI, NIRS having high-intensity data recording protocols, greater analysis of the subjects, and better treatment of the convalescents. Due to its speedy development and extensive implementation, neuroimaging has provided a completely different scenario in clinical neuroscience as well as in research associated with better cognition and brain dynamics.

REFERENCES

Abhang, P. A., Gawali, B., & Mehrotra, S. C. (2016). *Introduction to EEG-and speech-based emotion recognition.* Academic Press. https://books.google.com/books?hl=en&lr=&id=o4t4CgAAQBAJ&oi=fnd&pg=PP1&dq=introduction+to+eeg+and+speech+based&ots=pDinWkxU5l&sig=Exyz8LZpITWcqBEZzIFCzO7jUMc

Ashtaputre-Sisode, A. (2016). Emotions and brain waves. *The International Journal of Indian Psychology, 3*(2), 14–18. https://books.google.com/books?hl=en&lr=&id=XniVCwAAQBAJ&oi=fnd&pg=PA14&dq=EMOTIONS+AND+BRAIN+WAVES&ots=NjAQTRCOTY&sig=UoOC2RNvKKuoEW3FBEZp5YhdtdM

Bandettini, P. A. (2009). What's new in neuroimaging methods? *Annals of the New York Academy of Sciences, 1156*(1), 260–293. 10.1111/j.1749-6632.2009.04420.x

Biasiucci, A., Franceschiello, B., & Murray, M. M. (2019). Electroencephalography. *Current Biology, 29*(3), R80–R85. 10.1016/j.cub.2018.11.052

Braboszcz, C., Cahn, B. R., Levy, J., Fernandez, M., & Delorme, A. (2017). Increased gamma brainwave amplitude compared to control in three different meditation traditions. *PloS one, 12*(1), e0170647. 10.1371/journal.pone.0170647

Buskila, Y., Bellot-Saez, A., & Morley, J. W. (2019). Generating brain waves, the power of astrocytes. *Frontiers in neuroscience, 13*, 1125. 10.3389/fnins.2019.01125

Chen, J. E., & Glover, G. H. (2015). Functional magnetic resonance imaging methods. *Neuropsychology Review, 25*(3), 289–313. 10.1007/s11065-015-9294-9

Chen, W. L., Wagner, J., Heugel, N., Sugar, J., Lee, Y. W., Conant, L., ... & Whelan, H. T. (2020). Functional near-infrared spectroscopy and its clinical application in the field of neuroscience: advances and future directions. *Frontiers in Neuroscience*, *14*, 724. 10.3389/fnins.2020.00724

Cho, J. H., Lee, H. K., Dong, K. R., Kim, H. J., Kim, Y. S., Cho, M. S., & Chung, W. K. (2011). A study of alpha brain wave characteristics from MRI scanning in patients with anxiety disorder. *Journal of the Korean Physical Society*, *59*(4), 2861–2868. https://www.researchgate.net/profile/Woon-Kwan-Chung/publication/270110634_A_Study_of_Alpha_Brain_Wave_Characteristics_from_MRI_Scanning_in_Patients_with_Anxiety_Disorder/links/570df7f808aed31341cf87f0/A-Study-of-Alpha-Brain-Wave-Characteristics-from-MRI-Scanning-in-Patients-with-Anxiety-Disorder.pdf

Desai, R., Tailor, A., & Bhatt, T. (2015). Effects of yoga on brain waves and structural activation: A review. *Complementary therapies in clinical practice*, *21*(2), 112–118. 10.1016/j.ctcp.2015.02.002

Drevets, W. C. (2000). Neuroimaging studies of mood disorders. *Biological Psychiatry*, *48*(8), 813–829. 10.1016/S0006-3223(00)01020-9

Dudeja, J. P. (2017). Scientific analysis of mantra-based meditation and its beneficial effects: An overview. *International Journal of Advanced Scientific Technologies in Engineering and Management Sciences*, *3*(6), 21–26. https://www.researchgate.net/profile/Jai-Dudeja/publication/318395933_Scientific_Analysis_of_Mantra-Based_Meditation_and_its_Beneficial_Effects_An_Overview/links/5baa003192851ca9ed23aabd/Scientific-Analysis-of-Mantra-Based-Meditation-and-its-Beneficial-Effects-An-Overview.pdf

Ehlis, A. C., Schneider, S., Dresler, T., & Fallgatter, A. J. (2014). Application of functional near-infrared spectroscopy in psychiatry. *Neuroimage*, *85*, 478–488. 10.1016/j.neuroimage.2013.03.067

Glover, G. H. (2011). Overview of functional magnetic resonance imaging. *Neurosurgery Clinics*, *22*(2), 133–139. 10.1016/j.nec.2010.11.001

Goriely, A., Budday, S., & Kuhl, E. (2015). Neuromechanics: from neurons to brain. *Advances in Applied Mechanics*, *48*, 79–139. 10.1016/bs.aams.2015.10.002

Gui, X. U. E., Chuansheng, C. H. E. N., Zhong-Lin, L. U., & Qi, D. O. N. G. (2010). Brain imaging techniques and their applications in decision-making research. *Xin li Xue Bao. Acta Psychologica Sinica*, *42*(1), 120. 10.3724%2FSP.J.1041.2010.00120

Irani, F., Platek, S. M., Bunce, S., Ruocco, A. C., & Chute, D. (2007). Functional near infrared spectroscopy (fNIRS): an emerging neuroimaging technology with important applications for the study of brain disorders. *The Clinical Neuropsychologist*, *21*(1), 9–37. 10.1080/13854040600910018

Ismail, W. W., Hanif, M., Mohamed, S. B., Hamzah, N., & Rizman, Z. I. (2016). Human emotion detection via brain waves study by using electroencephalogram (EEG). *International Journal on Advanced Science, Engineering and Information Technology*, *6*(6), 1005–1011. https://www.researchgate.net/profile/Wan-Woaswi/publication/329584955_Human_Emotion_Detection_Via_Brain_Waves_Study_by_Using_Electroencephalogram_EEG/links/5c10b7a64585157ac1bbb648/Human-Emotion-Detection-Via-Brain-Waves-Study-by-Using-Electroencephalogram-EEG.pdf

Izzetoglu, M., Izzetoglu, K., Bunce, S., Ayaz, H., Devaraj, A., Onaral, B., & Pourrezaei, K. (2005). Functional near-infrared neuroimaging. *IEEE Transactions on Neural Systems and Rehabilitation Engineering*, *13*(2), 153–159. 10.1109/TNSRE.2005.847377

Jebelli, H., Hwang, S., & Lee, S. (2018). EEG signal-processing framework to obtain high-quality brain waves from an off-the-shelf wearable EEG device. *J. Comput. Civ. Eng*, *32*(1), 04017070. https://www.researchgate.net/profile/Houtan-Jebelli/publication/319550134_An_EEG_Signal_Processing_Framework_to_Obtain_High-Quality_Brain_Waves_from_an_Off-the-Shelf_Wearable_EEG_Device/links/5c85a1cd92851c69506b1fa8/An-EEG-Signal-Processing-Framework-to-Obtain-High-Quality-Brain-Waves-from-an-Off-the-Shelf-Wearable-EEG-Device.pdf

Jeong, E. G., Moon, B., & Lee, Y. H. (2011). A platform for real time brain-waves analysis system. In *International Conference on Grid and Distributed Computing* (pp. 431–437). Springer, Berlin, Heidelberg. 10.1007/978-3-642-27180-9_53

Jia, X., & Kohn, A. (2011). Gamma rhythms in the brain. *PLoS Biology*, *9*(4), e1001045. 10.1371/journal.pbio.1001045

Koudelková, Z., Strmiska, M., & Jašek, R. (2018). Analysis of brain waves according to their frequency. *Int. J. Of Biol. And Biomed. Eng.*, *12*, 202–207. https://www.researchgate.net/profile/Zuzana-Koudelkova-4/publication/334805116_Analysis_of_brain_waves_according_to_their_frequency/links/5d4194e84585153e59312c60/Analysis-of-brain-waves-according-to-their-frequency.pdf

Kringelbach, M. L., & Deco, G. (2020). Brain states and transitions: insights from computational neuroscience. *Cell Reports*, *32*(10), 108128. 10.1016/j.celrep.2020.108128

Matthews, P. M., & Jezzard, P. (2004). Functional magnetic resonance imaging. *Journal of Neurology, Neurosurgery & Psychiatry*, *75*(1), 6–12. https://jnnp.bmj.com/content/75/1/6.short

Müller-Putz, G. R. (2020). Electroencephalography. *Handbook of Clinical Neurology*, *168*, 249–262. 10.1016/B978-0-444-63934-9.00018-4

Noggle, C. A., & Davis, A. S. (2021). Advances in neuroimaging. In *Understanding the Biological Basis of Behavior* (pp. 107–137). Springer, Cham. 10.1007/978-3-030-59162-5_5

Proudfoot, M., Woolrich, M. W., Nobre, A. C., & Turner, M. R. (2014). Magnetoencephalography. *Practical neurology*, *14*(5), 336–343. 10.1136/practneurol-2013-000768

Raichle, M. E. (1983). Positron emission tomography. *Annual Review of Neuroscience*, *6*(1), 249–267. https://www.annualreviews.org/doi/pdf/10.1146/annurev.ne.06.030183.001341

Rice, L., & Bisdas, S. (2017). The diagnostic value of FDG and amyloid PET in Alzheimer's disease—A systematic review. *European Journal of Radiology*, *94*, 16–24. 10.1016/j.ejrad.2017.07.014

Savoy, R. L. (2001). History and future directions of human brain mapping and functional neuroimaging. *Acta Psychologica*, *107*(1-3), 9–42. 10.1016/S0001-6918(01)00018-X

Schacter, D. L. (1977). EEG theta waves and psychological phenomena: A review and analysis. *Biological Psychology*, *5*(1), 47–82. 10.1016/0301-0511(77)90028-X

Sejnowski, T. J., Koch, C., & Churchland, P. S. (1988). Computational neuroscience. *Science*, *241*(4871), 1299–1306. 10.1126/science.3045969

Singh, S. P. (2014). Magnetoencephalography: basic principles. *Annals of Indian Academy of Neurology*, *17*(Suppl 1), S107. 10.4103%2F0972-2327.128676

Stinson, B., & Arthur, D. (2013). A novel EEG for alpha brain state training, neurobiofeedback and behavior change. *Complementary Therapies in Clinical Practice*, *19*(3), 114–118. 10.1016/j.ctcp.2013.03.003

Supek, S., & Aine, C. J. (2016). *Magnetoencephalography*. Springer-Verlag Berlin An. https://link.springer.com/content/pdf/10.1007/978-3-642-33045-2.pdf

Teplan, M. (2002). Fundamentals of EEG measurement. *Measurement Science Review*, *2*(2), 1–11. http://www.edumed.org.br/cursos/neurociencia/Methods EEGMeasurement.pdf

Walker, K. A., & Bilgel, M. (2021). PET Neuroimaging of Neurologic Disease: Methods, Clinical and Research Applications. 10.31234/osf.io/u72e8

Wilson, T. W., Heinrichs-Graham, E., Proskovec, A. L., & McDermott, T. J. (2016). Neuroimaging with magnetoencephalography: A dynamic view of brain pathophysiology. *Translational Research*, *175*, 17–36. 10.1016/j.trsl.2016.01.007

Zimmer, L. (2009). Positron emission tomography neuroimaging for a better understanding of the biology of ADHD. *Neuropharmacology*, *57*(7-8), 601–607. 10.1016/j.neuropharm.2009.08.001

WEB SOURCE

https://imotions.com/wp-content/uploads/2022/10/brain-lobes-iMotions.png
https://assets.nhs.uk/nhsuk-cms/images/E5RF5X.width-1534.jpg
https://ars.els-cdn.com/content/image/3-s2.0-B9780128044902000026-f02-01-9780128044902.jpg
https://ccnmtl.columbia.edu/projects/neuroethics/module1/foundationtext/index.html
https://imotions.com/blog/learning/research-fundamentals/eeg-vs-mri-vs-fmri-differences/
https://imotions.com/wp-content/uploads/2022/10/fMRI-explained.jpg
https://info.tmsi.com/blog/types-of-eeg-electrodes
https://www.york.ac.uk/psychology/research/york-neuroimaging-centre/research/magnetoencephalography/
https://radiology.ucsf.edu/patient-care/services/specialty-imaging/alzheimer
https://www.pnas.org/doi/10.1073/pnas.2208729119

Chapter 5

EEG

Concepts, research-based analytics, and applications

Rashmi Gupta, Sonu Purohit, and Jeetendra Kumar
Atal Bihari Vajpayee University, Bilaspur, Chhattisgarh, India

5.1 INTRODUCTION

EEG is an abbreviation for electroencephalogram. It is a medical test that measures brain electrical activity. The electrodes are attached to the scalp and pick up the electrical signals produced by the brain, which are then transmitted to a machine, which displays the results in the form of a graph. The EEG is used to diagnose a wide range of neurological conditions, including epilepsy, sleep disorders, and brain damage. EEG devices record electrical signals generated from the brain. "Electroencephalography (EEG) has been instrumental in making discoveries about cognition, brain function, and dysfunction" [1]. It is already known that the human brain is always generating electrical signals, even when the person is sleeping. During different types of activities, the brain generates many waves. The signals are also known as neuro signals. In EEG, metal discs (electrodes) pick up the electrical activity that is generated by the neurons inside the brain cells. In the recording process of EEG, tiny metal discs (electrodes) are connected to the scalp to capture the electrical activity of the brain. The sensors carry out the signals received by a machine and display them on the computer monitor. The signals are in the form of waves. The recording procedure is done by a specialist, who is called a neurologist, in the hospital for a diagnostic test or any other medical purpose research center for research purposes. For the computerized processing of these brain signals generated from EEG devices, at first, these signals are amplified and then transferred to the computer. After preprocessing, machine learning and deep learning based algorithms are used for data analysis and making predictions. This process is shown in Figure 5.1.

Designing technology of EEG: EEG technology involves the use of electrodes attached to the scalp to pick up electrical signals produced by the brain, amplifiers to increase the strength of the signals, filters to remove any unwanted noise, and an analog-to-digital converter to convert the signals into a digital format that can be analyzed and displayed. The EEG machine is typically made up of a computer and specialized software

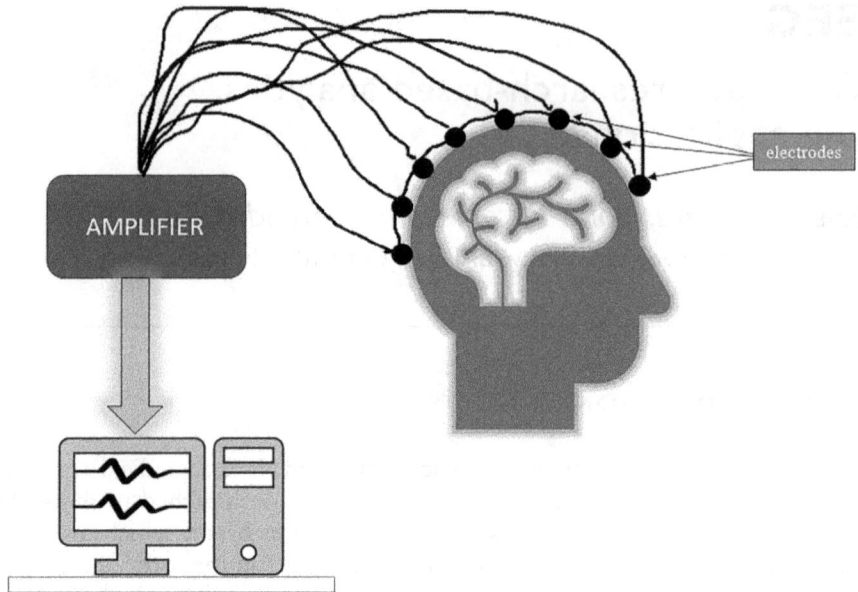

Figure 5.1 EEG signal processing.

that processes the data and displays the results in a visual format, such as a graph or image. Over the years, advances in materials, electronics, and software have resulted in smaller, more portable, and more efficient machines with higher resolution and accuracy. Furthermore, advances in brain-computer interface technology have enabled the development of EEG-based systems that can be used for a variety of applications other than traditional medical diagnosis, such as controlling prosthetic devices, monitoring brain activity for cognitive assessment and research, and more. EEG is a subpart of the brain-computer interface(BCI). BCI devices can be of the following types:

A. **Invasive:** In these types of devices, microelectrodes are directly placed into the cortex. These types of EEG devices are generally inserted into the internal area of the brain using surgical procedures. These invasive BCI devices effectively read brain signals, because they are placed inside the brain.

B. **Semi-invasive:** In these types of devices, electrodes are placed on the upper surface of the brain using a surgical procedure. These types of devices are known as ECoG (electrocorticography).

C. **Non-invasive:** In these types of devices, sensors are placed on the scalp of the head. This type of device does not very effectively read the signals, but it is used mostly because it does not require a surgical procedure to read brain waves as well as there is no chance of physical harm to the brain as compared with invasive and semi-invasive devices. These types of devices are known as EEG or MEG devices.

EEG devices can also be wired or wireless. The wired brain-wear devices offer strong signals and more stability than the wireless ones. There is no chance of dysconnectivity during the recording of electrical activity. During the utility of wired connected devices, we face a lack of movement but it is easy to record without any external interruption. In the wireless headset, the recording can be affected by connectivity. Due to the mobility of the device, it is easy to handle and can be used from a distance. The primary drawback of the wireless device is that sometimes the device can be disconnected for external reasons. Battery backup of the device can be one of the reasons for an interruption during recording.

5.2 PREPROCESSING TECHNIQUES OF EEG SIGNALS

Electric signals inside the brain may be affected by different types of reasons like body movements, eye blinking, opening the mouth, moving the leg as well as the imagination. Many variations can be seen in recorded signals when a person moves her/his leg, hand, head, or another part of the body. This kind of variation can be called noise or artifacts. The removal of noise from the datasets is called the preprocessing of datasets. Two types of artifacts can be present in the dataset – the first one is an internal artifact, and the second one is a system artifact. The internal artifact can arise due to body movements of the subject like eye blinking, hand, foot, or head movements, and system artifacts can arise due to fluctuations in the power supply in the device, movements in the electrodes of the device, etc.

There are a few artifact removal techniques to remove artifacts from the raw EEG dataset that are called filtering. The filtering includes three techniques, which are the high pass filter, low pass filter, and notch filter. The high pass filters out the slow frequencies that are less than 0.1 hz or 1.0 hz. High-pass filtering is used for removing drift and trends. The low pass filter is used for removing unwanted noise from data and is to make the dataset clear and improve its quality. The low pass filter removes high-frequency interferences greater than 50-70hz. EEG signals are often affected by power line noise, and such noise is sometimes avoided from a source with a carefully equipped system, but this work cannot always be successful. Moreover, it cannot work in the already gathered dataset. Therefore, notch filters are used for the removal of such noises from the dataset. The notch filter removes 50/60 Hz interference in the dataset. Apart from all these basic techniques, some new EEG data processing techniques are also proposed in some research works. Following are some of these techniques.

5.3 MACHINE LEARNING AND DEEP LEARNING BASED EEG DATA ANALYSIS TECHNIQUES

The EEG data consists of the activity of billions of neurons. Most of the time, we want to classify or analyze the EEG data according to some activity. For example, we can classify the EEG data according to the attention state of students during learning for mental disease classification, sleep stage classification, controlling the robotic action, etc. For the classification task, many artificial intelligence (AI) methods are used. In broad categories, these AI methods can be machine learning methods and deep learning methods. Machine learning techniques can be defined as computer-based models and algorithms that can learn from the datasets. There are so many machine learning techniques for data analysis.

Nowadays deep learning techniques are widely used in various areas. In the field of EEG analysis, deep learning techniques are very useful. In the last few years, some of the deep learning techniques have been very successful in the analysis of complex datasets such as EEG, audio, and image signals. Deep learning, a subfield of machine learning technology, helps to find out the feature extraction of the dataset. It can produce a higher level of performance for the different analysis tasks. Many deep learning techniques like RNN (recurrent neural network), LSTM (long short-term memory), DBN (deep belief network), and CNN (convolution neural network) are being used for EEG data analysis and classification, and some of the recent research work is shown in Table 5.1.

5.4 APPLICATIONS OF EEG

EEG devices are able to grasp the internal state of mind using the electrical signals generated from the mind. EEG is used to measure the electrical activity of the brain in a non-invasive manner. Because this activity is related to a variety of functions such as movement, sensation, thought, and emotion, EEG is a useful tool for understanding brain function. Furthermore, EEG can be used to diagnose various neurological disorders as well as monitor brain activity during surgery, which is critical for patient safety. EEG is also useful for researching sleep and assessing the efficacy of brain stimulation techniques. Furthermore, EEG's ability to detect changes in brain activity during drug trials is critical for the development of effective medicines for a variety of diseases. Due to the non-invasive nature of EEG, it can also be used by a non-medical person. Even people related to computers can also record brain waves and analyze them. In this chapter, EEG applications in some areas like neuro-marketing, behavioral science, cognitive neuroscience,

Table 5.1 Deep learning-based EEG data analysis techniques

Research work	Objective	Year	Dataset used	No. of volunteers	AI technique used	Model used	Accuracy
[2]	To classify schizophrenia during resting state	2023	Own dataset	31	Machine learning	KNN LR DT RF SVM	Highest 89% accuracy with SVM
[3]	To classify emotional valence	2023	DEAP dataset [4]	32	Machine learning	SVM Classifier	97.42% accuracy
[5]	To detect drowsiness	2023	Public dataset by Jianliang Min et al [6]	12	Machine learning	NB SVM KNN RFA	100% highest accuracy
[7]	To classify bipolar and other depressive disorders	2023	Own collected dataset	57	Machine learning	SVM KNN RF	89.3% accuracy
[8]	To diagnose brain disorders	2019	The TUH EEG Corpus [9]	13500	Deep learning	HMM (Hidden Markov Model)	90%
[10]	To find the efficient classification method for different frequency data	2021	Andrzejak et al. [11]	24	Deep learning	CNN-E	CNN-E model performed well with same frequency sampling and different frequency sampling
[12]	To classify eye close and eye-opening states using deep learning	2022	Own dataset	27	Deep learning	DLVQ (deep learning vector quantizer)	91% F-Score

(Continued)

Table 5.1 (Continued) Deep learning-based EEG data analysis techniques

Research work	Objective	Year	Dataset used	No. of volunteers	AI technique used	Model used	Accuracy
[13]	Epileptic patient classification	2022	Bonn EEG Database Bern-Barcelona Database	5 10	Deep learning	Attention-based wavelet CNN	98.89%
[14]	To classify motor imagery task	2022	Own collected dataset	57	Deep learning	CSP+LDA 2D CNN	Average 69.07% F- score
[15]	To classify emotions	2022	DEAP dataset [4]	32	Deep learning	Bi-LSTM	Accuracy 99.45%, 96.67%, and 99.68% for valence, arousal, and liking, respectively
[16]	To classify raw EEG	2022	PhysioNet EEG motor movement/imagery dataset [17]	109	Deep learning	Transformer based approach	Highest 83.31% accuracy
[18]	To detect schizophrenia using sound perception	2022	One public dataset and self-collected dataset	128	Deep learning	CNN	78% accuracy
[19]	To classify normal and alcoholic people	2022	UCI dataset	122	Deep learning	DWT+ CNN+ Bi-LSTM	99.32%
[20]	To predict tinnitus treatment outcomes	2023	Own collected dataset	9	Deep learning	TFI score + CNN	Accuracy ranging from 98%-100%
[21]	To predict awakening from the coma	2023	Own collected dataset	145	Deep learning	CNN	Positive prediction - 0.83 ± 0.03 and Negative prediction - 0.57 ± 0.04

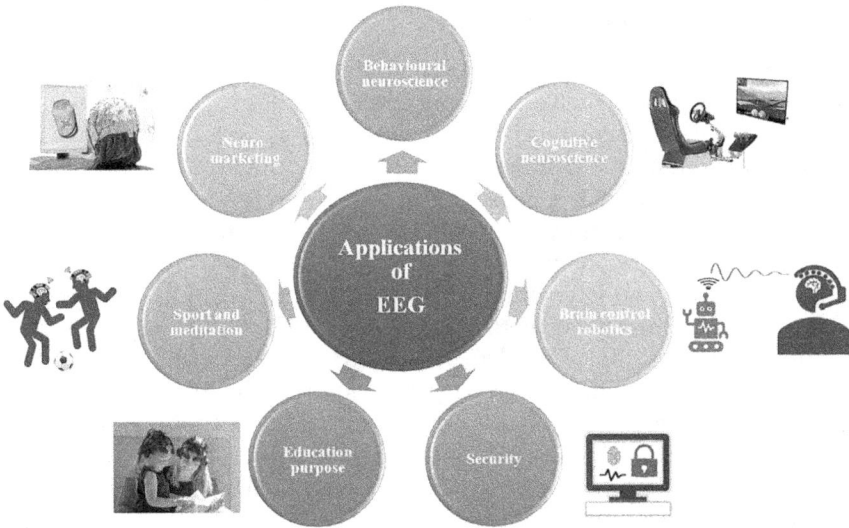

Figure 5.2 Applications of EEG.

education, security, etc. have been discussed. Figure 5.2 illustrates the applications of EEG in many areas.

5.4.1 Cognitive neuroscience

Cognitive neuroscience is the study of the brain's relationship to behavior. It combines psychology, neuroscience, and computer science to better understand how the brain processes information and generates behavior. Cognitive neuroscientists study brain activity using techniques such as EEG, MRI, and functional MRI (fMRI) to better understand the underlying mechanisms of cognitive processes such as attention, memory, language, decision making, and perception. Cognitive neuroscience is an inter-disciplinary field that draws knowledge from many disciplines, including biology, physics, and mathematics. The ultimate goal of cognitive neuro-science is to shed light on the neural basis of mental processes and to develop new treatments for neurological and psychiatric conditions. By understanding the relationship between brain activity and behavior, cog-nitive neuroscientists can improve our understanding of the human mind and develop more effective treatments for conditions such as Alzheimer's disease, depression, and schizophrenia. The field of cognitive neuroscience is highly interdisciplinary and draws on knowledge from a wide range of fields, including biology, physics, mathematics, and computer science. This interdisciplinary approach allows cognitive neuroscientists to tackle com-plex questions about the brain and behavior with a variety of tools and techniques, leading to discoveries and advancements in the field.

5.4.2 Behavioral neuroscience

The study of the biological and physiological basis of behavior is known as behavioral neuroscience. It combines psychology, neuroscience, and biology to better understand how the nervous system controls behavior and how genetics, development, and the environment influence behavior. Behavioral neuroscience is also known as bio-psychology. This is a study of behavior development in humans as well as animals. In this study, the psychological events are examined through the neurons by the biological activity. Animal models, brain imaging (e.g., EEG, MRI, and fMRI), and genetic and molecular methods are used by behavioral neuroscientists to study behavior. Behavioral neuroscience seeks to identify the neural circuits and mechanisms that underpin various behaviors such as learning and memory, emotion, sensation, movement, and perception. Behavioral neuroscientists can improve our understanding of the brain and its functions, as well as develop new treatments for neurological and psychiatric conditions, by understanding the biological basis of behavior. The field of behavioral neuroscience also investigates the effects of drugs and other substances on behavior, as well as the neural mechanisms underlying these effects.

5.4.3 Neuro-marketing

The application of neuroscience techniques to the study of consumer behavior and decision making is known as neuro-marketing. It seeks to comprehend how people make purchasing decisions and what factors influence them. Neuro-marketing measures brain activity and responses to marketing stimuli using tools such as EEG, fMRI, and eye-tracking. Neuro-marketing researchers can gain insights into consumer preferences, attitudes, and motivations by studying brain activity. These insights can then be used to develop more effective marketing strategies. Because it is still a young field, neuro-marketing's methods and conclusions can be debatable. However, by offering a deeper understanding of consumer behavior and the underlying neural mechanisms that drive it, it has the potential to completely alter how businesses approach marketing and consumer research. Neuro-marketing is a highly interdisciplinary field that draws on knowledge from marketing, psychology, and neuroscience. It is also a field that is constantly changing as new strategies and tactics are created. For neuro-marketing, EEG is an easy-to-use device.

Benefits of EEG in Neuro-marketing:
  i. **Analysis of consumer's decision-making process:** With the help of EEG signals, the decision-making process that is running in the brain of the consumer can be analyzed. Moreover, the buying behavior of the consumer can be found.

ii. **Analysis of marketing behavior:** In this method, the analysis process can be done using EEG signals. Analyzing consumer behavior in marketing strategies and analysis of brain functioning of liking/disliking any product can be done. This process can be a little investigation to find out the interest of the customer in buying any product to improve marketing goals.

iii. **Setup the marketing goals:** The main objective of this process is the brainstorming of the consumer without any inquiries. Spread a commercial message to the consumer and develop their interest to purchase the product by hidden extracting patterns with the help of EEG.

5.4.4 Sports and meditation

Meditation sessions can change the way of thinking. The stress can be decreased by meditation or sports. The changes can be seen in neurons inside the human brain in the form of several types of frequencies like delta, beta, theta, and alpha by the EEG technique. The amplification of alpha frequency, even in the non-meditative closed eyes state, is the EEG sign of meditation. EEG is used for a variety of reasons in both sports and meditation. To enhance performance and lower the risk of injury in sports, EEG can be used to monitor brain activity while an athlete is engaged in physical activity. An athlete's cognitive state can be tracked using EEG, which can also be used to spot signs of fatigue, stress, and changes in attention that could affect performance. Then, with this knowledge, training programs can be improved, recovery can be accelerated, and injuries can be avoided. EEG is used in meditation to track changes in brain activity as practitioners progress. EEG can be used to study the effects of meditation, which have been shown to have a variety of advantages, including lowered stress, enhanced mood, and increased focus. EEG can also be used for comparing different meditation practices.

5.4.5 Educational purpose

EEG can be applied in academic settings to examine and enhance the learning process. A student's brain activity during academic tasks like reading, problem solving, and memory exercises can be studied using EEG, which measures the electrical activity of the brain. By monitoring these brainwave signals in real time, it's easy to find out how the students are feeling during the study in their minds. If a teacher gives them a task, then he can see how they respond along with the feelings that are running inside their brain. The teachers can monitor their students' engagement level during learning by analyzing the brain waves with the EEG process. So, they can know the state of the student's brain, how the student received the task, whether the task was easy or too hard to understand, or whether the mind of the student was going through stress.

5.4.6 Security

User authentication system plays a very important role in security. There are many types of user authentication systems like pin, password, fingerprint, face recognition systems, etc. A new authentication system is going to be introduced named EEG authentication system. EEG contains a large number of psychological activities running in the human brain. There are many differences between an individual's brain structure and cognitive activities. The EEG signals of different people are different. But the same individual's brain performs the same activities, which are even repeatable. The same frequency of the brain can be used as authentication by EEG signals for security purposes. The EEG-based authentication system is best in cyber security or cryptography as it is unique always and not replicable as well. The EEG-based biometric system needs the BCI application to develop security systems.

5.4.7 Brain control robotics

Electrodes are affixed to the scalp to record the electrical activity of the brain in brain control technology. The signals from this activity are then processed and interpreted by a computer, which employs algorithms to convert them into control instructions for the robot. Numerous potential uses for brain-controlled robots include assistive technology for the disabled, rehabilitation technology for those with motor impairments, and research tools for examining brain-machine interactions. Additionally, brain-controlled robots have the potential to completely change the field of robotics by allowing humans to control robots directly with their brains. An EEG-based brain control robotic is a gadget or a robot that receives a command from a human operator and performs different activities according to that. The performance of brain control devices depends on different BCIs, which may have different speeds and accuracies as well. Table 5.2 shows recent research work done in many application areas of EEG.

5.5 CHALLENGES ASSOCIATED WITH EEG

EEG is a device that measures the electrical activity of the brain during any mental task. From the first viewpoint, anyone can say that it is technology to access the brain state of humans. New research doors are already open for conducting research in neuroscience using EEG devices, but as with other technologies, many challenges are also associated with EEG. These challenges can be categorized as follows.

Table 5.2 Latest research work done on different application areas of EEG

EEG application area	Research work & objective	Year	Dataset used	Device used	No of. volunteer	Methodology used	Findings
Cognitive Neuroscience	[22] Diagnosis of Alzheimer's disease	2023	Publically available dataset [23]	Walter EEGPL-2311	59	SVM	Accuracy 97.22%
	[24] To predict the decision-making process	2021	Own collected dataset	HydroCel Geodesic Sensory Net	74	Statistical analysis	EEG signals can discriminate decision variables.
	[25] To predict decision making	2020	Own collected dataset	Brain Products GmbH)	14	COH	Highest accuracy 0.90 ± 0.10
Behavior Neuroscience	[26] To analyze the effect of colors, graphics, and design	2022	Own collected dataset	BioSemi Inc.,	81	Statistical analysis	Way-finding cognition was significantly improved with color and graphic enhancement.
	[27] To detect driver distraction	2022	Own collected dataset	-	6	Bi-LSTM	Highest accuracy 92.48%
	[28] To analyze brain behavior and synchrony during parent–child interactions	2021	Own collected dataset	BioSemi B.V.	186	Statistical analysis	72% of sessions were completed.

(Continued)

Table 5.2 (Continued) Latest research work done on different application areas of EEG

EEG application area	Research work & objective	Year	Dataset used	Device used	No of. volunteer	Methodology used	Findings
Neuro-marketing	[29] To classify preferences	2022	DEAP dataset [4]	Biosemi Active Two system	32	DNN RF SVM KNN	Highest accuracy 94%
	[30] To predict consumer's emotion	2022	Publicly available dataset [31]	Emotiv Epoc+	25	Ensemble classification using GA	Accuracy 96.89%
	[32] To classify consumer preference	2022	Own collected dataset	Neurowerk EEG Sigma	45	K means clustering	Highest accuracy 92.4%
Sports and Meditation	[33] To access mental fatigue and stress	2021	Own collected dataset	Muse EEG Device	10	VAS ruler	Game positively affected stress and concentration
	[34] To access the mental status of rifle shooter	2019	Own collected dataset	Mindwave and neurosky	14	Calculated average of shooting score, attention, and heart rate before 5 s of shooting	Shooting score and meditation were higher on the expert, and attention and heart rate were higher on the novice.
Education	[35] To analyze split attention effects	2022	Own collected dataset	g.Nautilus, g.tec	40	Statistical test	The split-attention group in the beta brain wave and the focused attention group in the alpha brain wave both showed a significant difference.

		Year	Dataset	Device	No.	Method	Result
	[36] To analyze attention	2020	Own collected dataset	Mindwave	13	Comparison of attention values	Different media of study differently affects attention value.
Security	[37] To develop EEG-based personal authentication system	2022	Own collected dataset	EGI GES 300	15	Auto Weka	95.6%
	[38] To analyze EEG-based authentication	2019	Own collected dataset	Truscan EEG device	20	KNN LDA	Accuracy 98.04%
Brain Control Robotics	[39] To develop brain control robotic arm	2020	Own collected dataset	–	15	MDCBN	Success rate 0.60
	[40] To decode hand movements	2020	Own collected dataset	EEG with 57 electrodes	15	LDA	Average 48% score

5.5.1 Technical challenges

- The placement of electrodes on the scalp and their stability during recording can significantly impact the quality of the EEG signal. Any movement or displacement of electrodes can cause artifacts in the EEG data.
- The EEG signal's quality can be impacted by the electrical impedance of electrodes, which can vary depending on several factors like skin preparation, sweat, and oils.
- Every time, a special gel is required to be applied to electrodes during data collection. In the case of dry electrodes, data contain more noise and artifacts than wet electrodes.
- Poor fitting of the EEG cap can occur on different head sizes.
- High synchronization between the amplifier and EEG device is required.
- The choice of reference electrodes and their placement can significantly impact the interpretation of the EEG data.
- The EEG signals can also be contaminated by electrical activity generated by the muscles, especially in the vicinity of electrodes. This is known as an electromyographic (EMG) artifact and can make it difficult to distinguish between brain and muscle activity.

5.5.2 Social and ethical challenges

- Some people are afraid of participating in research conducted with EEG.
- Wearing an EEG headset for a long time results in a headache.
- EEG data can be highly sensitive and personal, as it provides information about an individual's thoughts and mental processes. This raises privacy concerns, as EEG data could be used to profile individuals or access their personal information.
- The use of EEG data in research or clinical settings requires informed consent from the participants, which can be difficult to obtain in some cases, such as in children or individuals with cognitive or neurological disorders.
- EEG data can be easily manipulated or misinterpreted, leading to incorrect conclusions about an individual's brain activity or mental state. This can have serious consequences for individuals, such as false diagnoses or inappropriate treatment.

5.5.3 Environmental challenges

- Ambient light and noise can affect the quality of the EEG signal, as both can generate electrical noise that can interfere with the EEG signal.

- Changes in temperature and humidity can affect the impedance of the electrodes and the stability of the EEG signal.
- The electrode gel used to prepare the scalp for EEG recording can dry over time, leading to increased impedance and decreased signal quality.
- EEG signals can be contaminated by other electrical devices in the environment, such as cell phones, computers, or medical equipment, which can generate electrical noise that interferes with the EEG signal.
- Movement during EEG recording can cause significant artifacts in the EEG signal, making it difficult to accurately interpret the data.

5.6 CONCLUSION

EEG is a non-invasive method used to measure the electrical activity of the brain. It involves placing electrodes on the scalp to record brainwaves, which can be analyzed to understand brain function and diagnose neurological disorders. EEG research has led to the development of advanced analytics techniques for analyzing brain signals, such as time-frequency analysis and machine learning algorithms. Numerous research is going on to develop efficient algorithms for preprocessing and classification of EEG signals. Highly efficient deep learning algorithms have been introduced by researchers to EEG analytics. In this chapter, we have discussed recent research done in the area of algorithm development for EEG analytics. We have also discussed recent research done in many application areas of EEG like cognitive neuroscience, behavioral neuroscience, sport, meditation, education, neuromarketing, etc.

REFERENCES

[1] M. X. Cohen, "Where Does EEG Come From and What Does It Mean?," *Trends Neurosci.*, vol. 40, no. 4, pp. 208–218, Apr. 2017.

[2] J. Ruiz de Miras, A. J. Ibáñez-Molina, M. F. Soriano, and S. Iglesias-Parro, "Schizophrenia Classification Using Machine Learning on Resting State EEG Signal," *Biomed. Signal Process. Control*, vol. 79, p. 104233, Jan. 2023.

[3] L. Abdel-Hamid, "An Efficient Machine Learning-Based Emotional Valence Recognition Approach Towards Wearable EEG," *Sensors 2023, Vol. 23, Page 1255*, vol. 23, no. 3, p. 1255, Jan. 2023.

[4] S. Koelstra *et al.*, "DEAP: A Database for Emotion Analysis; Using Physiological Signals," *IEEE Trans. Affect. Comput.*, vol. 3, no. 1, pp. 18–31, Jan. 2012.

[5] I. A. Fouad, "A Robust and Efficient EEG-based Drowsiness Detection System Using Different Machine Learning Algorithms," *Ain Shams Eng. J.*, vol. 14, no. 3, p. 101895, Apr. 2023.

[6] J. Min, P. Wang, and J. Hu, "Driver Fatigue Detection through Multiple Entropy Fusion Analysis in an EEG-based System," *PLoS One*, vol. 12, no. 12, Dec. 2017.

[7] M. Ravan *et al.*, "Discriminating between Bipolar and Major Depressive Disorder Using a Machine Learning Approach and Resting-state EEG Data," *Clin. Neurophysiol.*, vol. 146, pp. 30–39, Feb. 2023.

[8] M. Golmohammadi, A. H. Harati Nejad Torbati, S. Lopez de Diego, I. Obeid, and J. Picone, "Automatic Analysis of EEGs Using Big Data and Hybrid Deep Learning Architectures," *Front. Hum. Neurosci.*, vol. 13, p. 76, Feb. 2019.

[9] "Temple University EEG Corpus - Downloads." [Online]. Available: https://isip.piconepress.com/projects/tuh_eeg/html/downloads.shtml. [Accessed: 10-Feb-2023].

[10] T. Wen, Y. Du, T. Pan, C. Huang, and Z. Zhang, "A Deep Learning-Based Classification Method for Different Frequency EEG Data," *Comput. Math. Methods Med.*, vol. 2021, 2021.

[11] R. G. Andrzejak, K. Lehnertz, F. Mormann, C. Rieke, P. David, and C. E. Elger, "Indications of Nonlinear Deterministic and Finite-Dimensional Structures in Time Series of Brain Electrical Activity: Dependence on Recording Region and Brain State," *Phys. Rev. E*, vol. 64, no. 6, p. 061907, Nov. 2001.

[12] F. Husham Almukhtar, A. Abbas Ajwad, A. S. Kamil, R. A. Jaleel, R. Adil Kamil, and S. Jalal Mosa, "Deep Learning Techniques for Pattern Recognition in EEG Audio Signal-Processing-Based Eye-Closed and Eye-Open Cases," *Electron. 2022, Vol. 11, Page 4029*, vol. 11, no. 23, p. 4029, Dec. 2022.

[13] Q. Xin, S. Hu, S. Liu, L. Zhao, and Y. D. Zhang, "An Attention-Based Wavelet Convolution Neural Network for Epilepsy EEG Classification," *IEEE Trans. Neural Syst. Rehabil. Eng.*, vol. 30, pp. 957–966, 2022.

[14] N. Tibrewal, N. Leeuwis, and M. Alimardani, "Classification of Motor Imagery EEG Using Deep Learning Increases Performance in Inefficient BCI Users," *PLoS One*, vol. 17, no. 7, p. e0268880, Jul. 2022.

[15] M. Algarni, F. Saeed, T. Al-Hadhrami, F. Ghabban, and M. Al-Sarem, "Deep Learning-Based Approach for Emotion Recognition Using Electroencephalography (EEG) Signals Using Bi-Directional Long Short-Term Memory (Bi-LSTM)," *Sensors 2022, Vol. 22, Page 2976*, vol. 22, no. 8, p. 2976, Apr. 2022.

[16] J. Xie *et al.*, "A Transformer-Based Approach Combining Deep Learning Network and Spatial-Temporal Information for Raw EEG Classification," *IEEE Trans. Neural Syst. Rehabil. Eng.*, vol. 30, pp. 2126–2136, 2022.

[17] "EEG Motor Movement/Imagery Dataset v1.0.0." [Online]. Available: https://physionet.org/content/eegmmidb/1.0.0/. [Accessed: 10-Feb-2023].

[18] C. Barros, B. Roach, J. M. Ford, A. P. Pinheiro, and C. A. Silva, "From Sound Perception to Automatic Detection of Schizophrenia: An EEG-Based Deep Learning Approach," *Front. Psychiatry*, vol. 12, p. 2659, Feb. 2022.

[19] E. Tamilia, L. Ricci, H. Li, and L. Wu, "EEG Classification of Normal and Alcoholic by Deep Learning," *Brain Sci. 2022, Vol. 12, Page 778*, vol. 12, no. 6, p. 778, Jun. 2022.

[20] M. Doborjeh *et al.*, "Prediction of Tinnitus Treatment Outcomes Based on EEG Sensors and TFI Score Using Deep Learning," *Sensors 2023, Vol. 23, Page 902*, vol. 23, no. 2, p. 902, Jan. 2023.

[21] F. M. Aellen *et al.*, "Auditory Stimulation and Deep Learning Predict Awakening From Coma after Cardiac Arrest," *Brain*, Jan. 2023.

[22] A. An Approach *et al.*, "An Approach toward Artificial Intelligence Alzheimer's Disease Diagnosis Using Brain Signals," *Diagnostics 2023, Vol. 13, Page 477*, vol. 13, no. 3, p. 477, Jan. 2023.

[23] M. Cejnek, O. Vysata, M. Valis, and I. Bukovsky, "Novelty Detection-based Approach for Alzheimer's Disease and Mild Cognitive Impairment Diagnosis from EEG," *Med. Biol. Eng. Comput.*, vol. 59, no. 11–12, pp. 2287–2296, Nov. 2021.

[24] Y. Yau, T. Hinault, M. Taylor, P. Cisek, L. K. Fellows, and A. Dagher, "Evidence and Urgency Related EEG Signals during Dynamic Decision-Making in Humans," *J. Neurosci.*, vol. 41, no. 26, pp. 5711–5722, Jun. 2021.

[25] Y. Si *et al.*, "Predicting Individual Decision-Making Responses Based on Single-Trial EEG," *Neuroimage*, vol. 206, p. 116333, Feb. 2020.

[26] S. Kalantari *et al.*, "Evaluating the Impacts of Color, Graphics, and Architectural Features on Wayfinding in Healthcare Settings Using EEG Data and Virtual Response Testing," *J. Environ. Psychol.*, vol. 79, p. 101744, Feb. 2022.

[27] X. Zuo, C. Zhang, F. Cong, J. Zhao, and T. Hamalainen, "Driver Distraction Detection Using Bidirectional Long Short-Term Network Based on Multiscale Entropy of EEG," *IEEE Trans. Intell. Transp. Syst.*, vol. 23, no. 10, pp. 19309–19322, Oct. 2022.

[28] E. S. Norton *et al.*, "Social EEG: A Novel Neurodevelopmental Approach to Studying Brain-Behavior Links and Brain-to-Brain Synchrony During Naturalistic Toddler–Parent Interactions," *Dev. Psychobiol.*, vol. 64, no. 3, p. e22240, Mar. 2022.

[29] M. Aldayel, M. Ykhlef, and A. Al-Nafjan, "Deep Learning for EEG-Based Preference Classification in Neuromarketing," *Appl. Sci. 2020, Vol. 10, Page 1525*, vol. 10, no. 4, p. 1525, Feb. 2020.

[30] S. M. A. Shah *et al.*, "An Ensemble Model for Consumer Emotion Prediction Using EEG Signals for Neuromarketing Applications," *Sensors 2022, Vol. 22, Page 9744*, vol. 22, no. 24, p. 9744, Dec. 2022.

[31] M. Yadava, P. Kumar, R. Saini, P. P. Roy, and D. Prosad Dogra, "Analysis of EEG signals and its application to neuromarketing," *Multimed. Tools Appl.*, vol. 76, no. 18, pp. 19087–19111, Sep. 2017.

[32] S. Raiesdana and M. Mousakhani, "An EEG-Based Neuromarketing Approach for Analyzing the Preference of an Electric Car," *Comput. Intell. Neurosci.*, vol. 2022, 2022.

[33] S. Gündoğdu, Ö. H. Çolak, E. A. Doğan, E. Gülbetekin, and Ö. Polat, "Assessment of Mental Fatigue and Stress on Electronic Sport Players with Data Fusion," *Med. Biol. Eng. Comput.*, vol. 59, no. 9, pp. 1691–1707, Sep. 2021.

[34] "The psychophysiological differences between expert and novice rifle shooters during the aiming period." [Online]. Available: https://cyberleninka.ru/article/n/the-psychophysiological-differences-between-expert-and-novice-rifle-shooters-during-the-aiming-period/viewer. [Accessed: 09-Feb-2023].

[35] D. Mutlu-Bayraktar, P. Ozel, F. Altindis, and B. Yilmaz, "Split-Attention Effects in Multimedia Learning Environments: Eye-Tracking and EEG Analysis," *Multimed. Tools Appl.*, vol. 81, no. 6, pp. 8259–8282, Mar. 2022.

[36] D. Ni, S. Wang, and G. Liu, "The EEG-Based Attention Analysis in Multimedia m-Learning," *Comput. Math. Methods Med.*, vol. 2020, 2020.

[37] C. Stergiadis, V. D. Kostaridou, S. Veloudis, D. Kazis, and M. A. Klados, "A Personalized User Authentication System Based on EEG Signals," *Sensors 2022, Vol. 22, Page 6929*, vol. 22, no. 18, p. 6929, Sep. 2022.

[38] T. Zhi Chin, A. Saidatul, and Z. Ibrahim, "Exploring EEG Based Authentication for Imaginary and Non-Imaginary Tasks Using Power Spectral Density Method," *IOP Conf. Ser. Mater. Sci. Eng.*, vol. 557, no. 1, p. 012031, Jun. 2019.

[39] J. H. Jeong, K. H. Shim, D. J. Kim, and S. W. Lee, "Brain-Controlled Robotic Arm System Based on Multi-Directional CNN-BiLSTM Network Using EEG Signals," *IEEE Trans. Neural Syst. Rehabil. Eng.*, vol. 28, no. 5, pp. 1226–1238, May 2020.

[40] A. Schwarz, M. K. Höller, J. Pereira, P. Ofner, P. Ofner, and G. R. Müller-Putz, "Decoding Hand Movements from Human EEG to Control a Robotic Arm in a Simulation Environment," *J. Neural Eng.*, vol. 17, no. 3, p. 036010, May 2020.

Chapter 6

Classification of gait signals for detection of neurodegenerative diseases using log energy entropy and ANN classifier

Prasanna J, S. Thomas George, and M.S.P Subathra

Dept. of Robotics Engineering, Karunya Institute of Technology and Sciences, Coimbatore, Tamil Nadu, India

6.1 INTRODUCTION

The prevalence of neurodegenerative disorders (NDDs) has increased worldwide, and they have a considerable negative impact on economical, developmental, and health fronts [1]. They may result in severe movement issues, such as tremors in the limbs, jaw, or face; stiffness; or slowness of movement [2]. Gait analysis is a crucial method for evaluating NDD, even though the gait anomaly as a deviation from walking may indicate distinct problem patterns [3]. Neurodegenerative illnesses may manifest when neurons are harmed or begin to deteriorate. A number of serious NDDs, such as Parkinson's disease (PD), Huntington's disease (HD), Alzheimer's disease (AD), and amyotrophic lateral sclerosis (ALS), are brought on by changes in these cells that cause abnormal functioning and progressive damage of the arrangement of the neurons, including death of neurons [4]. Gait fluctuations analysis is a crucial method for evaluating NDD, even if gait abnormalities as a divergence from walking gait fluctuation may reflect many problem patterns [5,6]. Stride length and cadence, two fundamental temporal-spatial gait metrics, are used as input features in [7] by Kamruzzaman to assess the gait of people with cerebral palsy. The gait information explains the various gait patterns that humans exhibit while walking [6]. For the examination of PD gait, the authors of [8] revealed various regression normalization techniques that took physical characteristics and self-selected speed into account. In [9], Wu used a statistical analysis of gait rhythm with the nonparametric Paren-window approach to evaluate the probability density functions of intervals of stride, stance, and swing periods with its sub-phases.

Each gait cycle consists of a series of ordered gait events that take place at particular gait parameters indicated with temporal locations. The metrics of stride, posture, and swing are used to analyze human gait patterns to identify aberrant neurodegeneration classes. As the earliest study based on gait fluctuations-based illness analysis, Hausdorff et al. [10–12] examined

DOI: 10.1201/9781003398066-6

the stride interval time series of the gait in participants with HD as well as the healthy older subjects in comparison to control subjects. Their findings demonstrated that compared to controls, subjects with HD and older subjects have greater random stride interval fluctuations. [5,13,14] are more studies for consideration on gait oscillations analysis intended for illness state analysis. Driven by their findings, Kamruzzaman applied the support vector machine method and two fundamental temporal-spatial gait characteristics (stride length and cadence) as input features to examine the gait of people with cerebral palsy [15]. A study that uses fluctuation analysis and frequency range distribution to get a fresh perspective on gait rhythm is found in [16]. A tensor decomposition model for higher-dimensional analysis in PD was put forth in [17] by the author [18] described numerous regression normalization techniques for PD gait analysis that took into account the patient's physical characteristics and self-selected speed. With the statistical investigation of gait rhythm, Wu employed a nonparametric Paren-window approach in [19] to evaluate the gait intervals. For the gait dynamics analysis for classifying NDDs, frequency range distribution [20], tensor decomposition [21], and texture-based images with fuzzy recurrence plots [22] were proposed. The improvement of fall prediction, treatment, and rehabilitation procedures may result from research on the dynamics of gait patterns in neurodegenerative illness to determine the severity. Thus, we proposed non-linear entropy for extracting optimum features for the detection of diseases. The proposed research could aid in the classification of NDDs and the analysis of gait variations.

Contributions:
- An entropy-based feature extraction technique is proposed for effective feature extraction.
- Statistical-based features such as minimum, maximum, mean, energy, and normalized energy were used for the classification.
- An artificial neural network (ANN) classifier is utilized to classify the task, such as HC vs PD, HC vs. HD, HC vs. ALS, and HC vs. NDD.
- The classification performance shows better results with the proposed approach. The framework of the proposed method is shown in Figure 6.1.

Outline of the Paper:
- Section 6.2 provides the framework of the proposed methodology and dataset information and classification.
- In Section 6.3, we illustrate the performance of the classification of a different task, and we discuss the achieved results. The comparison between the existing and proposed method is reviewed based on the classification performance.
- Finally, Section 6.4 concludes the article with future directions.

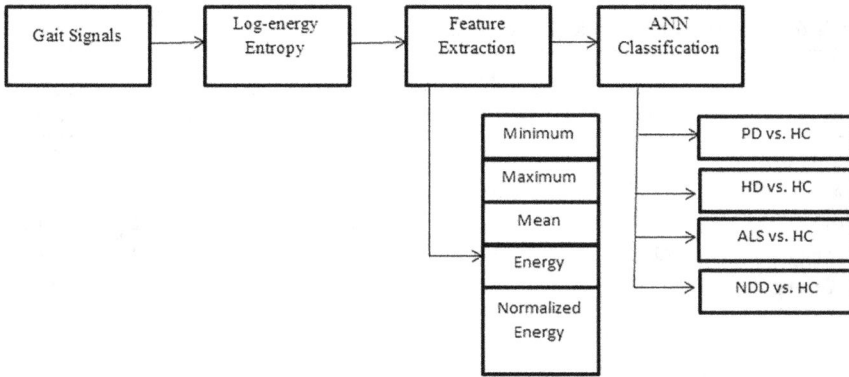

Figure 6.1 The framework of the proposed method.

6.2 METHOD AND MATERIALS

6.2.1 Dataset used

In this investigation, we use the Physionet [23] database's Gait Dynamics in Neuro-Degenerative Disease Dataset [24]. The dataset was suggested in order to comprehend the pathophysiology of NDDs better. There are 16 healthy controls, 15 PD patients, 20 HD patients, and 13 people with ALS. We abbreviate using HC, ALS, HD, and PD. The dataset includes a detailed description of the patients' clinical characteristics, including their age, gender, height, weight, walking speed, PD and HD illness severity, and ALS length. Examples of the HC, PD, HD, and ALS groups are shown in Tables 6.1–6.4, respectively.

Force-sensitive resistors, with an output roughly proportionate to the force under the foot, were used to collect the database's raw data. The signals were used to obtain the stride-to-stride measures of footfall contact periods, which include the left and right stride intervals, left and right swing intervals, left and right stance intervals, and double support intervals. Additionally, the percentages of stride for the left/right swing interval, left/right stance interval, and double support period were extracted. You can refer to [24,25] for a thorough description of the dataset's experiment conditions. Figure 6.2 displays examples for each category in the dataset.

Table 6.1 The classification performance using all five statistical features

| Classification task | Classification performance | | | | |
	Accuracy (%)	Sensitivity (%)	Specificity (%)	PPV (%)	NPV (%)
PD vs. HC	93.54	99.28	88.82	88.12	99.33
HD vs. HC	92.41	89.20	97.62	98.12	85.34
ALS vs. HC	91.10	84.97	97.89	97.50	86.00
NDD vs.HC	92.21	92.28	91.31	92.25	93.20

Table 6.2 Classification result (PD vs. HC)

Features	Classification performance (PD vs. HC)				
	Accuracy (%)	Sensitivity (%)	Specificity (%)	PPV (%)	NPV (%)
Minimum	90.17	89.18	90.71	91.22	89.22
Maximum	89.29	89.28	87.56	89.90	90
Mean	90.38	87.78	90.78	93.22	90.27
Energy	92.29	89.28	91.55	92.90	93.2
N-Energy	91.25	92.45	91.86	90.76	91.29

Table 6.3 Classification performance (HD vs. HC)

Features	Accuracy (%)	Sensitivity (%)	Specificity (%)	PPV (%)	NPV (%)
Minimum	91.72	90.68	90.97	91.92	87.22
Maximum	91.29	91.28	92.56	90.90	92.23
Mean	90.78	94.78	90.58	93.82	91.47
Energy	91.29	89	93.30	92.90	92
N-Energy	91.87	92.56	92.63	91.86	91

Table 6.4 Classification performance (ALS vs. HC)

Features	Accuracy (%)	Sensitivity (%)	Specificity (%)	PPV (%)	NPV (%)
Minimum	92.29	91.28	89.56	90.90	91.34
Maximum	90.38	87.78	90.78	93.22	90.27
Mean	91.72	90.68	90.97	91.92	87.22
Energy	91.25	92.45	91.86	90.76	91.29
N-Energy	90.17	89.18	90.71	91.22	89.22

6.2.2 Feature extraction

To effectively classify the gait patterns of NDDs (e.g., ALS, HD, and PD) and healthy people employed in this work, the standardized dataset is processed for the discovery of distinguishing feature sets. Statistical features have been computed by applying log energy entropy.

Log energy entropy is used to calculate the level of involvement in the non-stationary signals. Log energy entropy is defined mathematically as

$$\text{Log energy entropy} = \sum_{i=1}^{N} \log(y_i^2) \tag{6.1}$$

where N denotes the total length of the signal y_i and denotes the signal's ith sample. The log energy entropy feature was used by the authors of [26] and [27] to achieve high classification accuracy.

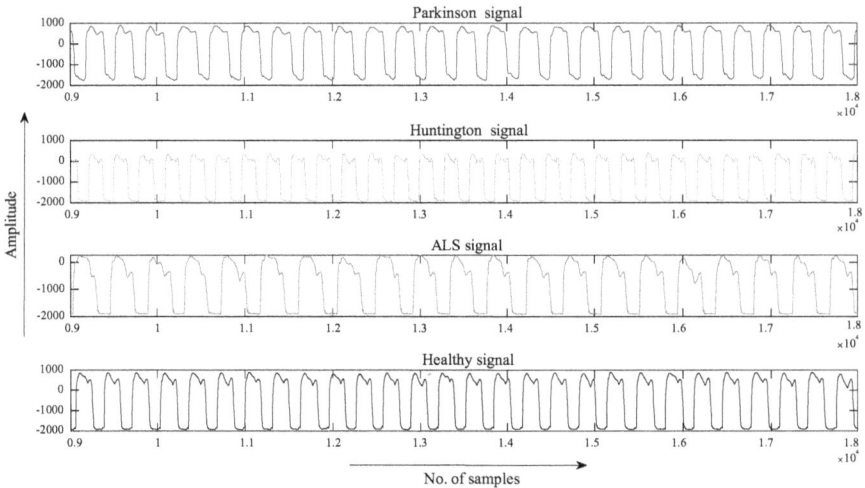

Figure 6.2 The visual representation of gait patterns of the different NDD classes.

The five optimum features – minimum, maximum, mean, energy, and normalized energy features – have been computed on both right and left foot.

6.2.3 Classification

Classification is a significant area of study in data mining, and neural networks are one of the most widely used techniques for classification. NDD classification employing gait features is typically challenged by three parameters that can affect performance: 1) a small number of clinical samples; 2) a large number of noisy or redundant features; and 3) the requirement to meet real-time requirements [28]. To eliminate these problems ANN classification is used. ANN is a complex adaptive system that can change its internal structure in response to the information it receives. It is accomplished by varying the weight of the connection. Each link carries a certain amount of weight.

A weight is a numerical value that governs the signal between two neurons. The output layer, input layer, and one or more hidden layers (middle layer) with a larger number of processing nodes or artificial neurons are all connected in an ANN [29]. The connection between two neurons is controlled by weights that can be adjusted to improve system accuracy, shown in Figure 6.3. Supervised learning is illustrated by an ANN. The knowledge was acquired by the ANN in the form of connected network units. Humans have a difficult time extracting this knowledge. This factor has motivated the extraction of classification rules in data mining.

ANN learning is achieved through the use of various training algorithms that are based on training guidelines or functions. Various training algorithms are used to recognize the training algorithm, which is based on

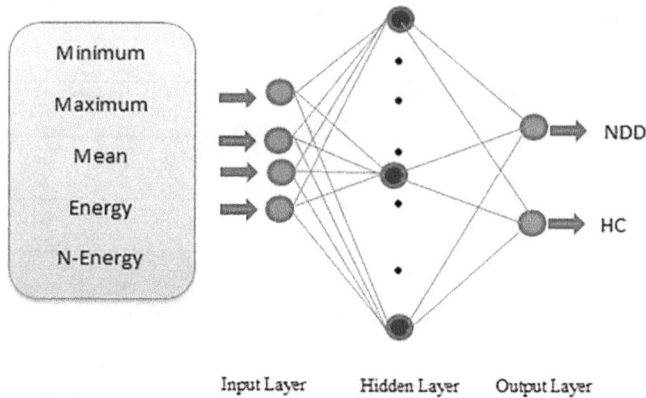

Figure 6.3 Functional unit of ANN classifier.

training rules or functions. The neural network is trained to perform its function by the training algorithm. In this work, the input layer is given five entropy features per channel, for a total of ten nonlinear features, to discriminate between normal healthy control and NDD gait signals. Physical modelling is commonly performed using a feed-forward multilayered neural network. The few hidden neurons used for classification are inefficient. As a result, the total number of hidden neurons studied in this study is ten. The neural network is trained using the Levenberg-Marquardt backpropagation (LMBP) training algorithm, which is a type of quasi-Newton algorithm.

6.2.3.1 Classification performance

The classification performance is computed for individual healthy control, PD, HD, and ALS subjects based on the classification results, such as classification accuracy, sensitivity, specificity, positive predictive value, and negative predictive value, which are given by the following formulas:

$$\text{Classification Accuracy} = \frac{True(+) + True(-)}{True(+) + True(-) + False(+) + False(-)} * 100 \quad (6.2)$$

$$\text{Sensitivity} = \frac{True(+)}{True(+) + False(-)} * 100 \quad (6.3)$$

$$\text{Specificity} = \frac{True(-)}{True(-) + False(+)} * 100 \quad (6.4)$$

$$\text{Positive predictive value (PPV)} = \frac{True(+)}{True(+) + False(+)} * 100 \quad (6.5)$$

$$\text{Negative Predictive Value(NPV)} = \frac{True\,(-)}{True\,(-)\,+\,False\,(-)} * 100 \qquad (6.6)$$

Where, True(+) indicates True positive, True(–) refers to True negative, False(+) corresponds to False positive and False(–) implies False negative.

6.3 RESULTS AND DISCUSSION

The gait signals taken from the pressure sensors from gait dynamics for the NDDs dataset is utilized to investigate the proposed technique. In the presented work, the non-linear characteristics of the gait signals are analyzed by applying log energy entropy computation on gait signals. It captures effective non-linear properties of the given signal where the optimum statistical-based features – minimum, maximum, mean, and energy and normalized energy – were captured from the left and right feet of a healthy control subject, an ALS patient, an HD patient, and a PD patient, respectively. The extracted features were given to the ANN classifier for further classification of the healthy control and individual with NDD. The total subjects taken for the experiment are 15 PD, 20 HD, 13 ALS, and 16 healthy controls. Since each recording has a 5-minute time duration with sampling of 300 samples for a single second, total number of samples for each foot is 90,000, respectively. The classification performance of the proposed work is shown in Table 6.1 and Figure 6.4.

The proposed method achieved the classification accuracy of 93.54% for the classification between PD vs. HC, 92.41% for HD vs. HC, 91.10% for ALS vs. HC, and 92.21% for NDD vs. HC. The sensitivity of PD vs. HC classification achieved 99.28% maximum sensitivity than all other classification tasks, but minimum specificity than all other classification tasks. Considering single statistical features for classification also performed

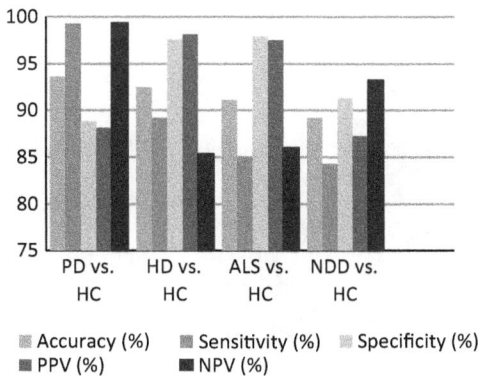

Figure 6.4 Classification results of proposed technique.

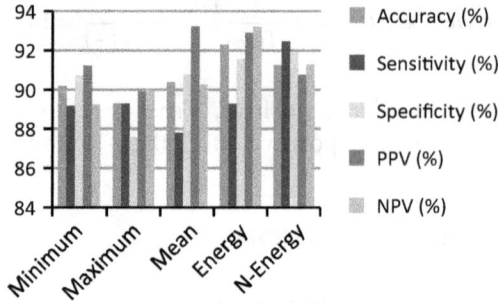

Figure 6.5 Classification performance of PD vs. HC.

better classification results. Table 6.2 and Figure 6.5 show the classification results of the classification between PD and HC by considering individual statistical features.

For the classification of PD vs. HC, while considering single features, energy features achieved the maximum classification performance with an accuracy of 92.29%. In the classification between HD vs. HC, normalized energy feature achieved the highest accuracy of 91.87% than all other features that are shown in Table 6.3 and Figure 6.6.

The classification result of ALS vs. HC is shown in Table 6.4 and Figure 6.7. Minimum feature outperforms the classification with an accuracy of 92.29%, sensitivity of 91.28%, specificity of 89.56%, PPV of 90.90%, and NPV of 91.34%, comparatively.

For the classification between the NDD and HC, all features have similar classification results with minor variations, shown in Table 6.5 and Figure 6.8. Hence, the proposed method produces promising results.

In comparison to some traditional variables for comparing multiple groups of subjects, the indices and HC appear to be far more useful. For example, if the p value is minimum of all Kruskal-Wallis tests for these subjects with disease, that are related to the statistical-like average value, standard deviation, fractal scaling index, decay time of autocorrelation, and nonstationarity

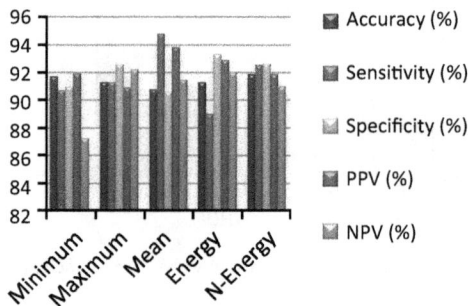

Figure 6.6 HD vs. HC results.

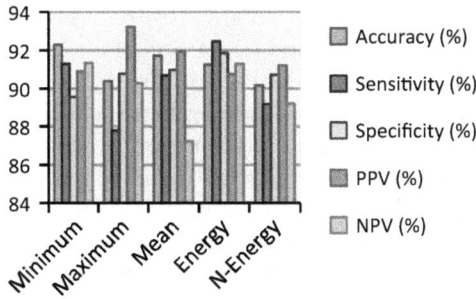

Figure 6.7 ALS vs. HC results.

Table 6.5 Classification performance (NDD vs. HC)

Features	Accuracy (%)	Sensitivity (%)	Specificity (%)	PPV (%)	NPV (%)
Minimum	91.92	90.56	90.27	91.62	89.22
Maximum	91.25	92.45	91.86	91.96	92.29
Mean	91.29	91.38	93.70	92.50	92.08
Energy	91.87	92.56	92.63	91.86	91.18
N-Energy	91.13	92.78	91.71	91.57	92.44

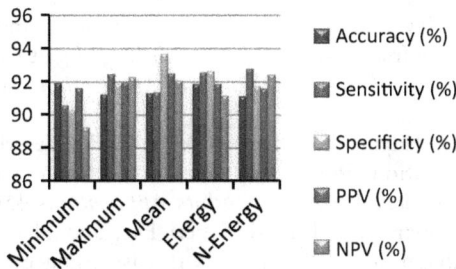

Figure 6.8 NDD vs. HC results.

index [20,24,30,31]. These findings clearly demonstrate the dependability of the proposed technique which may be significant for automatic diagnosing and continuous monitoring of all disease progression, that can be involved in the evaluation of potential therapy and treatment regimes. The proposed method may also assist physicians in making reliable decisions, that would eventually use individuals, families, society, and healthcare.

6.4 CONCLUSION

A new automated framework based on gait fluctuation measures was proposed in this study as a step forward into clinical and cost-effective detection of ALS, PD, and HD. To investigate a better metric for monitoring the progression of NDD and the effects of intervention therapies, the proposed entropy-based feature extraction using log energy entropy with statistical feature measures has produced a better classification of the entire task with promising results with consideration of single features and multiple features. This strongly demonstrates the dependability of classification results, which may be beneficial for automated intensive care of disease development as well as for early diagnosing. Furthermore, the proposed method may decrease clinical diagnosis imprecision. In the future, this research work will concentrate on improving the performance by incorporating deep learning techniques for classification.

REFERENCES

[1] Alaskar, Haya, Abir Jaafar Hussain, Wasiq Khan, Hissam Tawfik, Pip Trevorrow, Panos Liatsis, and Zohra Sbaï. "A data science approach for reliable classification of neuro-degenerative diseases using gait patterns." *Journal of Reliable Intelligent Environments* 6 (2020): 233–247.

[2] Jankovic, Joseph. "Parkinson's disease: clinical features and diagnosis." *Journal of Neurology, Neurosurgery & Psychiatry* 79, no. 4 (2008): 368–376.

[3] Hausdorff, Jeffrey M., Yosef Ashkenazy, Chang-K. Peng, Plamen Ch Ivanov, H. Eugene Stanley, and Ary L. Goldberger. "When human walking becomes random walking: fractal analysis and modeling of gait rhythm fluctuations." *Physica A: Statistical Mechanics and its Applications* 302, no. 1–4 (2001): 138–147.

[4] Lahmiri, Salim. "Image characterization by fractal descriptors in variational mode decomposition domain: application to brain magnetic resonance." *Physica A: Statistical Mechanics and its Applications* 456 (2016): 235–243.

[5] Shahbakhi, Mohammad, Danial Taheri Far, and Ehsan Tahami. "Speech analysis for diagnosis of Parkinson's disease using genetic algorithm and support vector machine." *Journal of Biomedical Science and Engineering* 2014 (2014).

[6] Salvatore, Christian, Antonio Cerasa, Isabella Castiglioni, F. Gallivanone, A. Augimeri, M. Lopez, G. Arabia, M. Morelli, M.C. Gilardi, and A. Quattrone. "Machine learning on brain MRI data for differential diagnosis of Parkinson's disease and Progressive Supranuclear Palsy." *Journal of Neuroscience Methods* 222 (2014): 230–237.

[7] Kamruzzaman, Joarder, and Rezaul K. Begg. "Support vector machines and other pattern recognition approaches to the diagnosis of cerebral palsy gait." *IEEE Transactions on Biomedical Engineering* 53, no. 12 (2006): 2479–2490.

[8] Wahid, Ferdous, Rezaul K. Begg, Chris J. Hass, Saman Halgamuge, and David C. Ackland. "Classification of Parkinson's disease gait using spatial-temporal gait features." *IEEE Journal of Biomedical and Health Informatics* 19, no. 6 (2015): 1794–1802.

[9] Wu, Yunfeng, and Sridhar Krishnan. "Statistical analysis of gait rhythm in patients with Parkinson's disease." *IEEE Transactions on Neural Systems and Rehabilitation Engineering* 18, no. 2 (2009): 150–158.

[10] Hass, Chris J., Thomas A. Buckley, Chris Pitsikoulis, and Ernest J. Barthelemy. "Progressive resistance training improves gait initiation in individuals with Parkinson's disease." *Gait & Posture* 35, no. 4 (2012): 669–673.

[11] McNeely, Marie E., and Gammon M. Earhart. "Medication and subthalamic nucleus deep brain stimulation similarly improve balance and complex gait in Parkinson disease." *Parkinsonism & Related Disorders* 19, no. 1 (2013): 86–91.

[12] Picelli, Alessandro, Camilla Melotti, Francesca Origano, Andreas Waldner, Antonio Fiaschi, Valter Santilli, and Nicola Smania. "Robot-assisted gait training in patients with Parkinson disease: a randomized controlled trial." *Neurorehabilitation and Neural Repair* 26, no. 4 (2012): 353–361.

[13] Eskofier, Bjoern M., Sunghoon Ivan Lee, Manuela Baron, André Simon, Christine F. Martindale, Heiko Gaßner, and Jochen Klucken. "An overview of smart shoes in the internet of health things: gait and mobility assessment in health promotion and disease monitoring." *Applied Sciences* 7, no. 10 (2017): 986.

[14] Genetics Home, Huntington disease. National Library of Medicine [Online]. https://ghr.nlm.nih.gov/condition/huntingtondisease(2018). Accessed 03 Jan 2018

[15] Kremer, H.P.H., and Hungtington Study Group. "Unified Huntington's disease rating scale: Reliability and consistency." *Movement disorders* 11 (1996): 136–142.

[16] Ren, Peng, Shanjiang Tang, Fang Fang, Lizhu Luo, Lei Xu, Maria L. Bringas-Vega, Dezhong Yao, Keith M. Kendrick, and Pedro A. Valdes-Sosa. "Gait rhythm fluctuation analysis for neurodegenerative diseases by empirical mode decomposition." *IEEE Transactions on Biomedical Engineering* 64, no. 1 (2016): 52–60.

[17] Pham, Tuan D., and Hong Yan. "Tensor decomposition of gait dynamics in Parkinson's disease." *IEEE Transactions on Biomedical Engineering* 65, no. 8 (2017): 1820–1827.

[18] Long, Jeffery D., Jane S. Paulsen, Karen Marder, Ying Zhang, Ji-In Kim, James A. Mills, and Researchers of the PREDICT-HD Huntington's Study Group. "Tracking motor impairments in the progression of Huntington's disease." *Movement Disorders* 29, no. 3 (2014): 311–319.

[19] Mannini, Andrea, Diana Trojaniello, Andrea Cereatti, and Angelo M. Sabatini. "A machine learning framework for gait classification using inertial sensors: Application to elderly, post-stroke and huntington's disease patients." *Sensors* 16, no. 1 (2016): 134.

[20] Hausdorff, Jeffrey M., Merit E. Cudkowicz, Renée Firtion, Jeanne Y. Wei, and Ary L. Goldberger. "Gait variability and basal ganglia disorders: Stride-to-stride variations of gait cycle timing in Parkinson's disease and Huntington's disease." *Movement disorders* 13, no. 3 (1998): 428–437.

[21] Barnéoud, Pascal, and Olivier Curet. "Beneficial effects of lysine acetylsalicylate, a soluble salt of aspirin, on motor performance in a transgenic model of amyotrophic lateral sclerosis." *Experimental Neurology* 155, no. 2 (1999): 243–251.

[22] Cho, Chien-Wen, Wen-Hung Chao, Sheng-Huang Lin, and You-Yin Chen. "A vision-based analysis system for gait recognition in patients with Parkinson's disease." *Expert Systems with Applications* 36, no. 3 (2009): 7033–7039.

[23] PhysioBank, PhysioToolkit. "Physionet: Components of a new research resource for complex physiologic signals." *Circulation* 101, no. 23 (2000): e215–e220.

[24] Hausdorff, Jeffrey M., Apinya Lertratanakul, Merit E. Cudkowicz, Amie L. Peterson, David Kaliton, and Ary L. Goldberger. "Dynamic markers of altered gait rhythm in amyotrophic lateral sclerosis." *Journal of Applied Physiology* (2000).

[25] Hausdorff, Jeffrey M., Susan L. Mitchell, Renee Firtion, Chung-Kang Peng, Merit E. Cudkowicz, Jeanne Y. Wei, and Ary L. Goldberger. "Altered fractal dynamics of gait: Reduced stride-interval correlations with aging and Huntington's disease." *Journal of Ppplied Physiology* 82, no. 1 (1997): 262–269.

[26] Saka, Kübra, Önder Aydemir, and Mehmet Öztürk. "Classification of EEG signals recorded during right/left hand movement imagery using Fast Walsh Hadamard Transform based features." In *2016 39th International Conference on Telecommunications and Signal Processing (TSP)*, pp. 413–416. IEEE, 2016.

[27] Gupta, Vipin, Tanvi Priya, Abhishek Kumar Yadav, Ram Bilas Pachori, and U. Rajendra Acharya. "Automated detection of focal EEG signals using features extracted from flexible analytic wavelet transform." *Pattern Recognition Letters* 94 (2017): 180–188.

[28] Sharma, Bhavna, and K.J.I.J.C.E. Venugopalan. "Comparison of neural network training functions for hematoma classification in brain CT images." *IOSR Journal of Computer Engineering* 16, no. 1 (2014): 31–35.

[29] Subasi, Abdulhamit, M. Kemal Kiymik, Ahmet Alkan, and Etem Koklukaya. "Neural network classification of EEG signals by using AR with MLE preprocessing for epileptic seizure detection." *Mathematical and Computational Applications* 10, no. 1 (2005): 57–70.

[30] Goetz, Christopher G., Werner Poewe, Olivier Rascol, Cristina Sampaio, Glenn T. Stebbins, Carl Counsell, Nir Giladi et al. "Movement Disorder Society Task Force report on the Hoehn and Yahr staging scale: Status and recommendations the Movement Disorder Society Task Force on rating scales for Parkinson's disease." *Movement Disorders* 19, no. 9 (2004): 1020–1028.

[31] Hausdorff, Jeffrey M. "Gait dynamics in Parkinson's disease: Common and distinct behavior among stride length, gait variability, and fractal-like scaling." *Chaos: An Interdisciplinary Journal of Nonlinear Science* 19, no. 2 (2009): 026113.

Chapter 7

An optimized text summarization for healthcare analytics using swarm intelligence

Rekha Jain[1], Pratistha Mathur[2], and Manisha[3]

[1]Department of Computer Applications, Manipal University Jaipur, India
[2]Department of Information Technology, Manipal University Jaipur, India
[3]Dept of Computer Science, Manipal University Jaipur, India

7.1 INTRODUCTION

Healthcare providers and persons involved in healthcare need accurate and updated information to make better decisions. As in healthcare, proper decision making can be life changing for patients and their relatives. Data analytics in healthcare, i.e., healthcare analytics, can play a vital role in this. Making decisions about surgery or therapy, predicting the course of major health events, and making long-term plans all depend on the decision makers' ability to promptly collect and analyze complete, reliable data. With digitalization, healthcare-related data and medical research documents are available in bulk. Therefore, users of the Internet do not wish to waste their valuable time on any material, whether it be a document, research paper, or book, without first understanding the significance of such material to their needs. Before spending their time reading the source papers, users like to have a summary or overview of such publications. Automatic text summarization (ATS) can play a vital role in this. Efficient summary generation has become challenging in this scenario. Swarm intelligence optimization methods can be used to generate an efficient summary. The following section discusses in detail text summarization, approaches of text summarization, the role of text summarization in healthcare, and the role of swarm intelligence in optimizing the summary(s). Section 7.2 focuses on a literature review of work done in healthcare analytics. Sections 7.3 and 7.4 describe the TF-IDF algorithm and PSO algorithm, respectively, that are used in the generation of an optimized summary. The proposed methodology is explained in section 7.5. Results are discussed in section 7.6. Finally, section 7.7 concludes the chapter with future work.

7.1.1 Text summarization

ATS techniques are becoming important as the amount of online text materials is increasing day by day. "A summary is a text produced

DOI: 10.1201/9781003398066-7

from one or more texts that contain a significant portion of the information in the original text(s) and is no longer than half of the original text(s)". Summaries can be categorized as indicative, informative, extractive, and abstractive. Indicative summaries give an idea of content, while informative summaries give a brief of the content. Content can be shortened by creating extracts or abstracts. Extracts are generated by reusing a portion of the text, i.e., sentences or words of input text, while abstracts are regenerated by using new phrases. Extractive text summarization and abstractive text summarization are the two primary methods of ATS.

7.1.2 Text summarization approaches

ATS systems look for the most important content contained in the original document(s). One can use the following two approaches.

7.1.2.1 Extractive text summarization

Extractive summarization techniques use scoring methods to assign an importance score or rank to each fragment of the text and return the highest scoring fragments to generate the summary.

7.1.2.2 Abstractive text summarization

Abstractive summarization compression or reformation methods are applied over extracted text. These summaries look like original texts and are more cohesive than extractive summaries.

Additionally, classification of summaries can be done as indicative or informative depending on the information they provide. The characteristics of indicative summaries are as follows:

- Without containing the substance of the original document, it provides hints regarding its contents.
- Its major goal is to indicate the content of the original document without divulging the specifics of the original text.
- It only provides a bare suggestion about the domain or primary subject of the input document.

However, an informative summary contains the following characteristics:

- It gives the document's key information.
- It presents the key information or concepts present in the original work, which means it must include all pertinent material from the original document and exclude additional details.

ATS may have a three-stage procedure. In the first stage, topic selection or identification, focus is on what portion of text to include in the summary. This step generally assigns scores to different portions of the text, which helps in the selection of the text. The second stage is topic interpretation, which performs fusion or compression and helps in briefing the content. The last or third stage is a summary generation, which produces the final summary in the desired form, and it uses mainly the text generation method to reformulate the text. Most of the existing systems use the first stage only and produce a summary using pure extracts.

Topic identification is based on assigning a score to each unit (e.g., word, clause, or sentence) of the input content and then producing the top scoring n-units as per required length of the summary. In literature, various methods are used for scoring the fragments of input text. Text summarization systems use an independent scoring method to score each unit of text.

To compute the score various criterion can be used and experimented in literature. The first criteria is positional criteria, which says that certain locations like headings, titles, first paragraphs, etc., contain important information, so these phrases are given a higher score. Second criterion is based on cue phrases in which sentences containing these phrases are given a higher score. A popular method of text identification or scoring is based on word and phrase frequently, which says that if a text has higher frequency of some words, then sentences containing those words are of higher importance and are given higher scores. If query-based summarization is needed, in that case, query and title overlap methods score the sentences high that contain desired words. The other criterion can be connectedness or discourse structure, which are finding the connectedness in the sentences.

Summaries generated by different approaches can be evaluated using different evaluation methods. Lin and Hovey [2003] introduced ROUGE to evaluate text summaries. ROUGE compares a system summary to a human summary using different methods like unigram, bigram, skip bigram, etc. ROUGE measures recall, i.e., number of system units included in the gold standard.

7.1.3 Text summarization in healthcare

The importance of ATS has been recognized in different verticals and domains. Healthcare and medicine is one among them. It helps medical practitioners and researchers to get more and precise information in a short span of time. The important sources of health information are electronic health records, clinical reports, and medical research publications [Rohil 2022]. These documents help healthcare professionals to learn about patients and their health conditions, and research articles help them to learn about new findings, therapies, and other advancements in healthcare. Abstractive summarization is generally used in literature to get concise information, whereas to get an in-depth overview, extractive summarization approaches are popularly implemented.

[Sarkar K 2009] worked in text summarization in the medical domain. In this work, they used domain knowledge to select important information from the input text of the medical domain. [Pivovarov R 2015] worked on the summarization of electronic health and clinical records. [Gayathri 2015] proposed a medical document summarization method using the extractive text summarization process, and topic selection was done using the cue phrase method. Medical Subject Heading (MeSH) was used to find cue phrases. [Milad et al.] and [Paul et al.] proposed deep learning methods also for text summarization in healthcare.

7.2 LITERATURE REVIEW

There are different types of medical records that can be useful for medical professionals for decision making such as electronic health records, clinical reports, medical research publications, etc. The summaries may help in better and quick understanding of the medical text data. In recent years, many researchers have done work related to summarization of health data (Table 7.1).

Table 7.1 Current research in text summarization for the health domain

S. No.	Author and publication	Technique used	Dataset
1.	Kanwal N and Rizzo G 2022	Used multi-head attention-based mechanism for performing extractive summarization	MIMIC-III discharge notes
2.	Jesus M et al. 2018	Multi Objective Artificial Bee Colony (MOABC)	Document understanding conference's dataset
3.	Gayathri P and Jaisankar N 2015	On the basis of cue word occurrences, important sentences were extracted from documents to get summarized text	For the identification of domain-specific terms, MeSH vocabulary thesaurus was used.
4.	Moradi M and Mattias S 2019	For biomedical text, summarization embeddings learned by BERT model were used	Retrieved articles from BioMed Central to construct development and evaluation corpora
5.	Deepika S et al. 2021	Pretrained models BERT, GPT-2 and Text Rank	The COVID-19 Open Research Dataset: CORD-19
6.	Kieuvongngam V et al. 2020	Pretrained models, BERT and OpenAI GPT-2	CORD-19

(Continued)

Table 7.1 (Continued) Current research in text summarization for the health domain

S. No.	Author and publication	Technique used	Dataset
7.	Gigioli P et al. 2018	Novel reinforcement learning reward metrics used for learning which is based on biomedical expert tools like UMLS Metathesaurus and MeSH	Biomedical literature from MEDLINE from NLM's PubMed citation database
8.	Chen YP et al. 2020	Bidirectional Encoder Representations using Transformers (BERT) based structure with a two-stage training method	A dataset of discharge diagnoses from NTUH-iMD
9.	Moradi M 2018	Clustering and Itemset mining-based Biomedical Summarizer	Self created text corpora
10.	Reddy SM and Mirivala S 2020	A multi-objective optimization approach with the use of similarity and position of the sentences. The value of similarity is calculated using TF-IDF and WMD.	Data from Mendeley datasets for summarization of clinical trial descriptions

[Kanwal N and Rizzo G 2022] presented an architecture that gave better results than frequency, graph-oriented, and centroid-based methods on MIMIC-III dataset. They suggested that if a concoction of abstractive and extractive approaches is used, it may give more reliable results.

[Jesus M et al. 2018] proposed an architecture using a multi-objective artificial bee colony method for multi-document summarization. They claimed it was the first-time implementation of the MOABC algorithm for text summarization. Results obtained showed improved ROUGE-2 score with other parallel methods.

[Gayathri P and Jaisankar N 2015] performed research and presented an approach for single medical document summarization. For this, domain-specific terms as cue words were used by model to fetch the important sentences from the dataset of documents.

[Moradi M and Mattias S 2019] used embeddings learned by BERT model for the summarization of biomedical text. For this, they randomly retrieved articles from BioMed Central. For the future, researchers suggested the use of contextual representation to address some problems, such as biomedical-named entity recognition, information extraction, etc.

[Deepika S et al. 2021] focused on cleaning the text as much as possible. After that, pretrained models GPT-2, Text Rank, and BERT were applied on the CORD-19 dataset. Out of these models, GPT-2 showed the best results for summarization.

[Kieuvongngam V et al. 2020] worked on the CORD-19 dataset and showed that the text-to-text multi-loss approach for training can be used for the fine tuning of a pre-trained model like GPT-2 to perform abstractive summarization. Results obtained were reasonable and interpretable.

[Gigioli P et al. 2018] proposed a deep reinforced summarization model that is capable of generating domain-based summaries of biomedical documents. They used reward metrics to boost the quality of generated summaries using a NLM PubMed citation dataset.

[Chen YP et al. 2020] presented a model named AlphaBERT model, which uses bidirectional encoder representation from a transformers-based structure with two-stage training. This approach decreases the size of the model without affecting the performance for summarization. This model helps staff of the hospital by giving quick summarized information.

[Moradi M. 2018] proposed clustering and itemset mining to develop a biomedical summarizer using self-created corpora. This model extracts concepts from the inputted data and uses itemset mining to extract main topics, and a clustering algorithm is used to place sentences in a cluster, which shows similar topics. After that, it generates summaries.

[Reddy SM and Miriyala S 2020] presented a multi-feature based optimization model for summarization, in which various features such as cosine similarity, position, and word-mover distance were used. Clinical trial description dataset was taken from Mendeley.

7.3 TF-IDF ALGORITHM

It is a statistical measure that finds how relevant a word is to a text document [Kim and Gil 2019]. The algorithm weighs keywords in the input text and assigns importance to each of them on the basis of the number of times they are present in the document. This is calculated with the help of two values. The first is term frequency, which means how many times a word is present in a document. This can be computed by counting the instances of the word in a document. The second is inverse document frequency of the keyword across a set of documents. It finds how rare or how common a word is in a set of documents [Manning et al. 2008]. If the value is closer to 0, the word is very common. The TF-IDF score is calculated by the formula given in Equation 7.1; here t is the keyword or term, d is the document, D is the complete document set, and N is the total number of documents present in the document set.

$$tf\_idf\ (term: t, document: d,) = TF\ (t, d) * IDF\ (t) \qquad (7.1)$$

Where:

$$TF(t, d) = \frac{Occurances\ of\ t\ in\ d}{Number\ of\ terms\ in\ d} \qquad (7.2)$$

$$IDF(t) = \log_e \frac{number\ of\ documents}{Number\ of\ documents\ with\ term\ t\ in\ it} \qquad (7.3)$$

This algorithm is used very commonly in machine learning, information retrieval, and text summarization/keyword extraction.

7.4 SWARM INTELLIGENCE USING PARTICLE SWARM OPTIMIZATION

Particle Swarm Optimization (PSO) is a simple algorithm for solving optimization problems. This algorithm is a simulation of social behaviors. [Reynolds] and [Heppner and Grenader] initially presented the simulation of bird flocking. PSO has roots in theory of bird flocking, fish schooling, and swarm theory. A sociobiologist E. O. Wilson wrote in reference to fish schooling: "In theory at least, individual members of the school can profit from discoveries and previous experiences of all other members of the school during the search of food." Kennedy and Eberhart proposed a swarm-based stochastic algorithm "Particle Swarm Optimization".

In PSO, each potential solution is assumed as one particle with some fixed velocity flying through the space of the problem. It assumes that knowledge is shared socially, not only between generations but also among members of the same generation. PSO optimizes a problem iteratively, starting with a set of solution candidates (population). These solution candidates are swarms of particles. Each particle knows the global best position within the swarm and its own best position found so far during the solution space search process. At every iteration, the velocity and the position of each particle in the swarm are influenced by individual and collective knowledge, which decides the next value in search of the optimal value. It tries to move towards a promising area to get the global optimal. The process is repeated until some stopping criterion is satisfied. The process can be summarized as follows:

In PSO, based on its prior flying experiences and those of its group-mates, each particle dynamically modifies the speed at which it travels. The best outcome for each particle is what is known as their personal best, or pbest. The global best, or gbest, is the best value for any particle, and each particle alters its position in accordance with its current position, current velocity, and the distance between its current position and p-best, its current location's separation from g-best.

7.4.1 Particle swarm optimization algorithm

Suppose F: objective function, A: population of agents, Vi: velocity of particle or agent, C1: cognitive constant, U1, U2: random numbers, C2: social constant, W: inertial weight Xi: position of particle or agent, p-best: Personal Best, g-best: Global Best

The PSO algorithm is as follows:

1. Generate the initial population of particles (agents) evenly distributed over X.
2. Evaluate the position of each particle taking the objective function $Z = F(x, y)$
3. Update p-best, i.e., If a particle's current position is better than its previous best position, update it.
4. Update g-best, i.e., Find the best particle (according to the last best places of particles).
5. Update the velocity of each particle using the formula given below.

$$V_i^{t+1} = W.\ V_i^t + c_1 U_1^t \left(P_{b_1}^t - P_i^t\right) + c_2 U_2^t (g_b^t - P_i^t) \qquad (7.4)$$

6. Update each particle position using the following formula:

$$P_i^{t+1} = P_i^t + v_i^{t+1} \qquad (7.5)$$

7. Check the stopping criterion; if false, go to step 2.

The most impressive aspect of PSO is its stable topology, which enables particles to communicate with one another and pick up new information more quickly in order to reach the global optimum. Because it optimizes a problem by constantly attempting to improve a potential solution, the metaheuristic aspect of this optimization algorithm gives us a lot of alternatives. Its application will grow as Ensemble Learning is studied more and more. Candidate summaries are viewed as particles in text summarizing. By applying the PSO, one concludes the summary(s), which can be considered as best as per the given criterion.

7.5 PROPOSED METHODOLOGY

7.5.1 Input text

A medical report of neurology is treated as input text, and it is considered for health analytics purposes. The dataset is taken from mtsamples.com, where various medical transcripts are available. The proposed methodology is shown below as a figure 7.1.

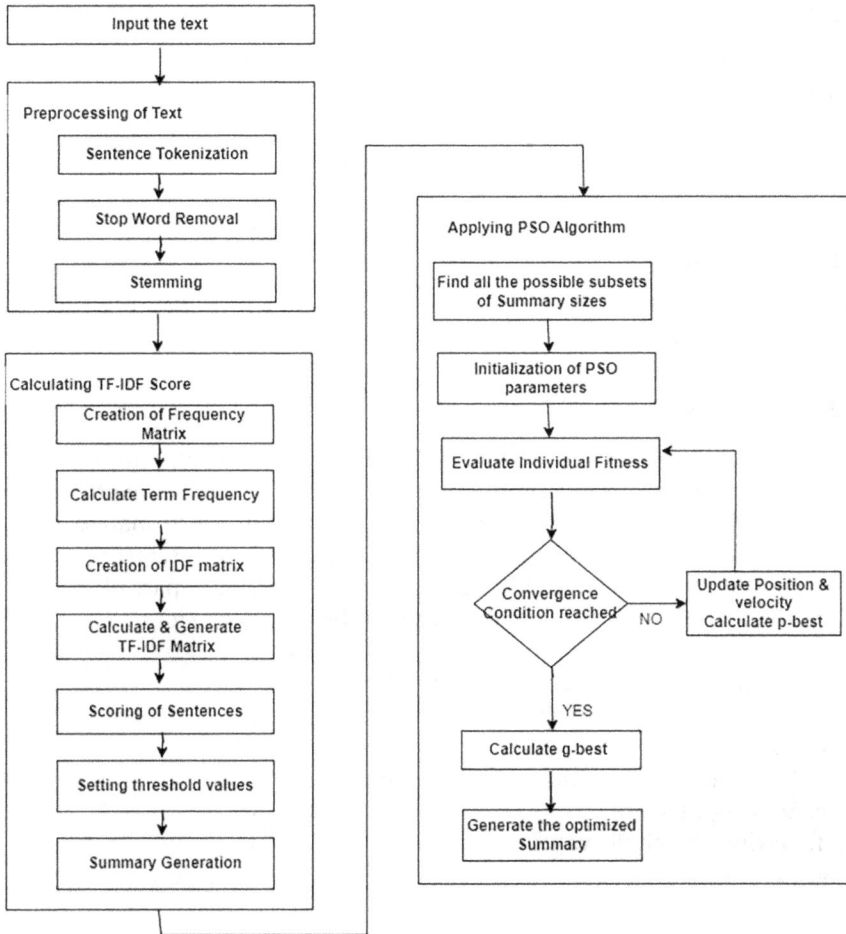

Figure 7.1 Proposed methodology.

7.5.2 Preprocessing

It is one of the most important steps because correctly preprocessed data give good results. If data are preprocessed in an incorrect manner, they will yield incorrect results. Finally, in preprocessing, we need to covert textual data into numeric form because machines cannot understand textual forms of data. They can only work with numeric data. Following are the steps involved in preprocessing of data.

7.5.2.1 Sentence tokenization

In this step, the entire text is broken down into number of sentences. Text is separated into sentences based on punctuation marks and space (.). Then each sentence is considered for further processing.

7.5.2.2 Stop word removal

Some words are there that do not carry meaning but are required to form a proper sentence. Examples of these words are "is," "am," "are," "has," "have," etc. These words are of no use for machines. They only reduce the importance of analysis by increasing the weight of text. Nltk toolkit is used to remove the stop words.

7.5.2.3 Stemming

In this step, all the words are reduced to their root forms. Porter's stemmer algorithm is used to find the root word.

7.5.3 Applying TF-IDF algorithm

Now the text is ready to feed to the TF-IDF algorithm. This algorithm is based on term frequency and inverse document frequency. It counts the frequency of words in each sentence and calculates the term frequency matrix. Then, it calculates the IDF matrix and finally provides the score to sentences. Based on some threshold value, only candidate sentences that are part of the summary are filtered. This algorithm consists of the following steps:

    a. Creation of a frequency matrix
    b. Calculate term frequency
    c. Creation of an IDF matrix
    d. Calculate & generate a TF-IDF matrix
    e. Scoring of sentences
    f. Setting threshold values
    g. Summary generation

7.5.4 Generation of different versions of summary

By iteratively running the algorithm and setting different threshold values, various versions of the summary are produced. Next, all possible versions are passed to the PSO algorithm.

7.5.5 Applying PSO algorithm

The PSO algorithm takes the initial score of generated summaries (particles) and their bounds. With the help of the fitness function, inertia (w), cognitive constant (c1), and social constant (c2), position and velocity of each particle are updated iterative until convergence is reached. The steps are as follows:

7.5.5.1 Find all possible sets of summaries

All the possible subsets of summaries generated by the TF-IDF algorithm are treated as particles.

7.5.5.2 Initialization of PSO parameters

Initialization is performed in this step. w, c1, c2, r1, r2 are initialized with some values. Bounds are also specified to ensure resultant values must fall within range.

7.5.5.3 Update the parameters until they get optimized or until some condition is reached

The PSO algorithm works in an iterative manner. The fitness value, position, and velocity of particles are updated iteratively until the values converge.

7.5.5.4 Get the p-best value for all versions

For each version, the p-best value is calculated by comparing the new p-best with the previous iteration's p-best value.

7.5.6 Evaluate the summaries and provide the best optimized summary as a result

The p-best values for all the particles are evaluated, and finally, the g-best value (best p-best value) is calculated. The summary that is associated with the g-best value is considered as the most optimized summarized result.

7.6 RESULTS AND DISCUSSIONS

The dataset is taken from Mtsamples.com. It belongs to medical specialty: Neurology. Sample Name is "Adult Hydrocephalus". This report contains information about a 74-year-old woman who needs a neurological consultation for possible adult hydrocephalus. She has mild gait impairment and mild cognitive slowing. For experimental purposes, the following medical transcription sample report is considered.

Original Text: "The patient is a lovely 74-year-old woman who presents with possible adult hydrocephalus. Danish is her native language, but she has been in the United States for many many years and speaks fluent English, as does her husband. With respect to her walking and balance, she states, "I think I walk funny". Her husband has noticed over the last six months or so that she has broadened her base and become more stooped in her pasture. Her balance has also gradually declined such that she frequently touches walls and furniture to stabilize herself. She has difficulty stepping up onto things like a scale because of this imbalance. She does not festinate. Her husband has noticed some slowing of her speed. She does not need to use an assistive device. She has occasional difficulty getting in and out of a car. Recently she has had more frequent falls. In March of 2007, she fell when she was walking to the bedroom and broke her wrist. Since that time, she has not had any emergency department trips, but she has had other falls. With respect to her bowel and bladder, she has no issues and no trouble with frequency or urgency. The patient does not have headaches. With respect to thinking and memory, she states she is still able to pay the bills, but over the last few months she states, "I do not feel as smart as I used to be". She feels that her thinking has slowed down. Her husband states that he has noticed, she will occasionally start a sentence and then not know what words to use as she is continuing. The patient has not had trouble with syncope. She has had past episodes of vertigo, but not recently. Significant for hypertension diagnosed in 2006, reflux in 2000, insomnia, but no snoring or apnea. She has been on Ambien, which has no longer been helpful. She has had arthritis since the year 2000, thyroid abnormalities diagnosed in 1968, a hysterectomy in 1986, and a right wrist operation after her fall in 2007 with a titanium plate and eight screws. Her father died of heart disease in his 60s, and her mother died of colon cancer. She has a sister who she believes is probably healthy. She has had two sons, one who died of a blood clot after having been a heavy smoker and another who is healthy. She had two normal vaginal deliveries. She lives with her husband. She is a nonsmoker and has no history of drug or alcohol abuse. She does drink two to three drinks daily. She completed 12th grade."

The mentioned summary is given as input to the proposed system, and various versions of the summary are produced. Each summary is treated as a particle, and the value of each particle is equivalent to the sum of sentence scores of the summary. These values of particles are passed to the PSO algorithm. This algorithm works in an iterative manner, and values of velocity, positions, and p-best are calculated until the convergence condition is reached. Finally, the g-best value is calculated among all the p-best values (Table 7.2).

Table 7.2 Summary and their p-best scores

Summary	30%	40%	45%	50%	55%	60%	65%	70%
p-best	2.48	3.24	2.95	3.92	3.07	4.08	4.64	4.73

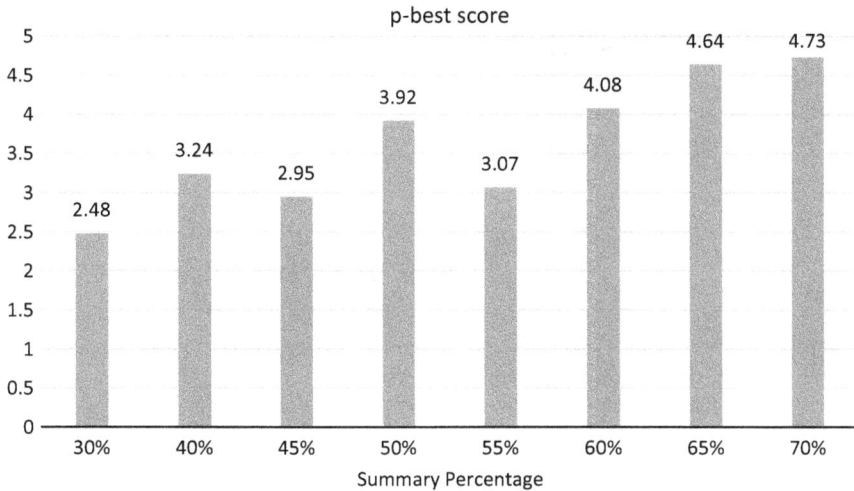

Figure 7.2 p-best score for various version of summary.

Values in the graph depict that the 70% summary is the best optimized summary. The g-best value is 4.73 as it is the best among all the p-best values.

7.7 CONCLUSION AND FUTURE WORK

In this study, the authors implemented extractive text summarization using the PSO approach for summarizing medical transcript reports that are available online. Initially, the TF-IDF algorithm is used to generate the sentence scores; then the PSO algorithm iteratively calculates the p-best values for each particle based on the distance from the summarized result. Finally, it calculates the g-best value that exhibits the optimal solution. In the future, the authors would work on abstractive text summarization and would try to generate more optimal solutions from abstractive summaries.

REFERENCES

Chen, Y.P., Chen, Y.Y., Lin, J.J., Huang, C.H., and Lai, F. 2020. Modified Bidirectional Encoder Representations From Transformers Extractive Summarization Model for Hospital Information Systems Based on Character-Level Tokens (AlphaBERT): Development and Performance Evaluation. *JMIR Med Inform.* doi: 10.2196/17787. PMID:30445218.

Deepika, S., Lakshmi Krishna, N., and Shridevi, S. 2021. Extractive Text Summarization for COVID-19 Medical Records. *Innovations in Power and Advanced Computing Technologies (i-PACT).* Kuala Lumpur. Malaysia. 1–5. doi: 10.1109/i-PACT52855.2021.9697019.

Gayathri, P., and Jaisankar, N. 2015. Towards an Efficient Approach for Automatic Medical Document Summarization. *Cybernetics and Information Technologies.* vol.15. no.4. 2015, 78–91. 10.1515/cait-2015-0056.

Gigioli, P., Sagar, N., Rao, A., and Voyles, J. 2018. Domain-Aware Abstractive Text Summarization for Medical Documents. *2018 IEEE International Conference on Bioinformatics and Biomedicine.* Doi: 10.1109/bibm.2018. 8621539.

Jesus, M., Sanchez-Gomez, M., Vega-Rodríguez, A., and Pérez, C.J. 2018. Extractive multi-document text summarization using a multi-objective artificial bee colony optimization approach. *Knowledge-Based Systems.* vol.159. 1–8. ISSN 0950-7051. 10.1016/j.knosys.2017.11.029.

Kanwal, N., and Rizzo, G. 2022. Attention-based clinical note summarization, SAC '22: Proceedings of the 37th ACM/SIGAPP Symposium on Applied Computing. (April): 813–820. 10.1145/3477314.3507256.

Kieuvongngam, V., Tan, B., and Niu, Y. 2020. Automatic text summarization of COVID-19 medical research articles using BERT and GPT-2. *CoRR.* Volume abs/2006.01997. http://dblp.org/rec/journals/corr/abs-2006-01977.bib.

Kim, S.W., and Gil, J.M. 2019. Research paper classification systems based on TF-IDF and LDA schemes. *Human-centric Computing and Information Sciences.* vol. 9. 30. 10.1186/s13673-019-0192-7

Manning C., Raghavan, P., and Schütze, H. 2008. *Introduction to Information Retrieval.* England: Cambridge University Press.

Moradi, M., and Samwald, M. 2019. Clustering of Deep Contextualized. for Summarization of Biomedical Texts. *arXiv.* doi: 10.48550/ARXIV.1908.02286.

Pivovarov, R., and Elhadad, N. 2015. Automated methods for the summarization of electronic health records. *J. Am. Med. Inf. Assoc.* vol.22. no. 5. 938–947.

Rohil, M.K., and Magotra, V. 2022. An exploratory study of automatic text summarization in biomedical and healthcare domain. *Healthcare Analytics.* vol. 2. 100058, ISSN 2772-4425, 10.1016/j.health.2022.100058.

Reddy, S.M., and Miriyala, S. 2020. Exploring Multi Feature Optimization for Summarizing Clinical Trial Descriptions. *IEEE Sixth International Conference on Multimedia Big Data (BigMM). New Delhi, Indi.* 341–345. Doi: 101109/ BigMM50055.2020.00059.

Sarkar, K. 2009. Using domain knowledge for text summarization in medical domain. *Int. J. Recent Trends Eng.* vol.1. no. 1. 200–205.

"Transcribed Medical Transcription Sample Reports and Examples." https://www.mtsamples.com/site/pages/sample.asp?Type=42-Neurology&Sample=1695-Adult%20Hydrocephalus.

Chapter 8

Computer aided diagnosis of neurodegenerative diseases using discrete wavelet transform and neural network for classification

Prasanna J[1], S. Thomas George[1], and M.S.P Subathra[2]
[1]Dept. of Biomedical Engineering, Karunya Institute of technology and sciences, Coimbatore, India
[2]Dept. of Robotics Engineering, Karunya Institute of technology and sciences, Coimbatore, Tamil Nadu, India

8.1 INTRODUCTION

Neurodegenerative diseases (NDDs) are defined by neurophysiological control as the enlightened damage of pattern or function of neurons, including neurodegeneration. Amyotrophic lateral sclerosis (ALS), Parkinson's disease (PD), and Huntington's disease (HD) are examples of common NDDs. Clinical symptoms of various NDDs may differ [1]. NDDs are a group of warning signs characterized by a decline in neural networks, cognitive abilities, and human motor systems. These disorders cause unusual changes in neuromuscular control [2]. Individuals with NDDs are at risk of sustaining serious physical injuries that will have an impact on their daily lives. This suggests the need for a precise diagnostic method capable of improving prognostication skills and assisting in the development of new endovascular effective treatments [3]. PD causes bradykinesia, rigidity, a resting tremor, and posture instability. The foremost clinical symptoms of HD are chorea and behavioral and cognitive changes, whereas the most visible symptom of ALS is muscular degenerative disorder and atrophy. Meanwhile, as a result of a malfunction in the neural central controller, the common external feature of NDD is neurological dysfunction, which has been investigated and confirmed according to several studies. PD and HD are both disorders of the basal ganglia that are characterized by changes in gait rhythm. ALS is an NDD that primarily affects the moto neurons of the prefrontal cortex, limbic system, and spinal cord [4].

Standard experimental procedures used to diagnose NDDs include biological blood tests, electromyography analysis [5], structural imaging practices, spinal cord imaging [6], and nerve biopsy [7]. These are, however, time-consuming, intrusive, or costly procedures, and some of these strategies may be less effective in the early stages of the diseases [8]. For decades, researchers have been studying the effects of these pathogens on

DOI: 10.1201/9781003398066-8

human locomotion [9]. There has been a lot of interest in automating the diagnosis of NDDs using gait dynamics [10], because it allows for a non-invasive and low-cost observation of the neuromuscular motor control development. Gait recordings are a complex combination of strength [11], sensation, and coordination. These physiognomies of the gait time series recommend that it may be sensitive to vascular dementia in mobility functioning and thus aid in the diagnosis of such illness [12]. The human gait is a series of cyclic spatial and temporal motions of the left and right feet that are directed through the nervous system [2]. Morris et al. [13] illustrated that PD had a significant impact on the subjects' gait by reducing their speed, stride frequency, and total range of walking motion. The reviewed studies found that stride interval or stride-to-stride fluctuation gait subtleties differed significantly between NDD patients and control subjects. Currently, the most frequently used method for diagnosing and evaluating advancement in NDD patients is static based on various types. It is well recognized that using questionnaires can result in independent results. As a result, an impartial evaluation of patients' physically efficient routine is critical in daily clinical practice, and it helps clinical doctors improve a more satisfactory treatment design and a more scientific analysis of therapy outcome [14]. Duta et al. [15] used Elman's recurrent neural network to identify healthy and unhealthy gait based on stance, swing, and double support intervals. Cross-correlograms of those gait signals with corresponding signals from a reference subject provided the relevant features in their work. In the context of the previous research, gait parameters related to the locomotor system's sensitivity to local perturbations, which is important for investigating muscular strength of balance. Detrended fluctuation analysis (DFA) was used by Hausdorff et al [16]. and Jordan et al. to show that gait dissimilarities are not casual but exhibit long-range correlation in healthy subjects. Furthermore, their research found that the scaling exponent values reflect locomotor control, adaptability, and versatility of walking in humans. The swing-interval turns count (SWITC) was used by Wu et al. to represent a significant variation between healthy control subjects and ALS patients [17]. Daliri et al. [18] extracted features such as the minimum, maximum, average, and standard deviation (SD) from each gait stride sequence and fed this relevant data to a support vector classifier to make a distinction of the healthy group from the NDD group. Wu and Krishnan [19] used the signal turns count procedure to assess gait variability with diseased patients. They discovered a significant difference in the swing interval turns count parameter between healthy control subjects and PD patients. This study [20] used the nonparametric frequency analysis based Parzen-Window utilized to determine the probability density functions (PDFs) of movement patterns from the stride interval, those can be realized the categorization of gait in subjects and normal subjects by calculating the average of the left-foot stride intermission and the altered Kullback-Leibler divergence from the PDF's projection.

In addition to significant work towards automatic gait analysis, the past few decades have seen the rapid growth of gait recognition, which has procedures similar to gait detection and classification. Moustakidis et al. [21] used wavelet packet (WP) degradation to measure ground reaction forces while walking, and then used a fuzzy complementary criterion to obtain a compact set of robust and corresponding features from WP coefficients using only one camera. Hu [22] anticipated a two-stage based Gaussian mixture model to depict the dual-tree complex wavelet transform (DTCWT) coefficients after applying gait images. A simple nearest-neighbor classifier was then used to measure the viewpoint of a gait sequence. Zeng and Wang [23] used the deterministic learning theory to extract features from a series of subject aesthetic and then discovered gait of DTCWT with different dimensions and positioning to gait energy cognition for the realization of inter gait recognition. A feature representation and a more appropriate classifier are required for improved performance. Difficulty stems from the following two fundamental facts: Due to the nonlinear dynamics of the human system, gait cadence exhibits complex and nonlinear behavior in both NDD and CO subjects [24]. The first 20 seconds of each and every recording were not considered in order to eliminate start-up consequences, and a median filter was used to remove sets of data that were three standard deviations above or below the median feature value. The cracks at the expiration of the hallway are also most likely responsible for these outliers; Hausdorff et al. proposed all of the preprocessing steps in their previous study [25]. In [25,26] NDD patients can be helped by using gait rhythm signals, which are easy to collect using various sensors. This study also focused on the detection of gait time series in order to describe their suitability in symmetry assessment for NDD detection. Classification method based on a machine learning approach was used to distinguish gait time intervals of NDD patients. Considering small number of training samples or noisy data, the proposed method is projected to attain higher classification performance for all classification tasks.

Contributions:

- Time-frequency domain based feature extraction technique is proposed for the effective feature extraction.
- Discrete wavelet transform (DWT) technique has been applied on the gait time series to decompose the signals to coefficients.
- Statistical and entropy-based features were extracted from the decomposed coefficients taken for the classification.
- Artificial neural network (ANN) classifier is utilized to classify the task such as HC vs PD, HC vs. HD, HC vs. ALS, and HC vs. NDD.
- The classification performance shows the better results with the proposed approach. The framework of the proposed method is shown in Figure 8.1.

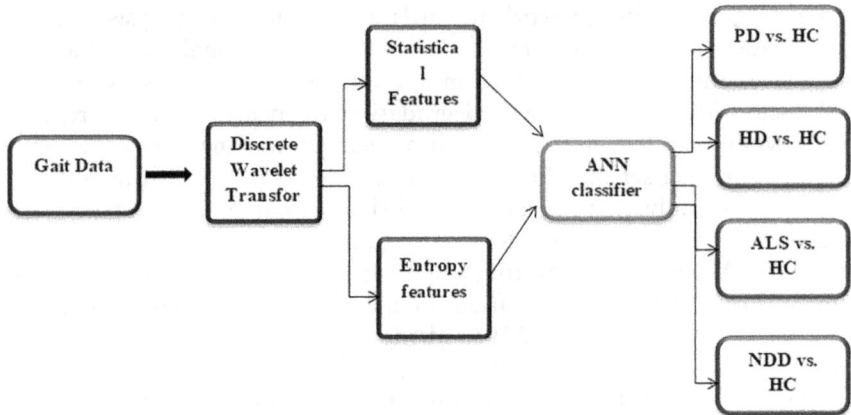

Figure 8.1 The pictorial framework of the proposed technique.

Outline of the Paper:

- Section 8.2 provides the framework of the proposed methodology and dataset description and classification technique.
- In Section 8.3, we illustrate the performance of the classification of different tasks, and we discuss the achieved results. The comparison between the existing and proposed method is reviewed based on the classification performance.
- Finally, Section 8.4 concludes the article with future directions.

8.2 METHODS AND MATERIALS

8.2.1 Dataset used

The gait dataset defined in [25] is used in this study. It is freely accessible at www.physionet.org. The whole gait recordings in the dataset include gait recordings from 20 HD subjects, 13 ALS subjects, 15 PD subjects, and 16 healthy control (HC) subjects. All patients with HD, PD, and ALS were evaluated for indicators that could impair gait signals. A comparison of HC and NDD subjects revealed no significant dissimilarities in heights and weights [26].

All participants were instructed to walk 77 meters in a straight line at their average speed. The gait signals were recorded using force-sensitive resistors located beneath each subject's feet for 5 minutes while walking [27,28].

The visual representation of gait signals is shown in Figure 8.2. A 12-bit analog-to-digital converter with a sampling rate of 300 Hz was used to collect the data. The data contains various time series of strike intervals defined as left and right stride intervals (Sec), swing intervals (Sec), stance intervals (Sec), and left and right stance intervals (% of stride).

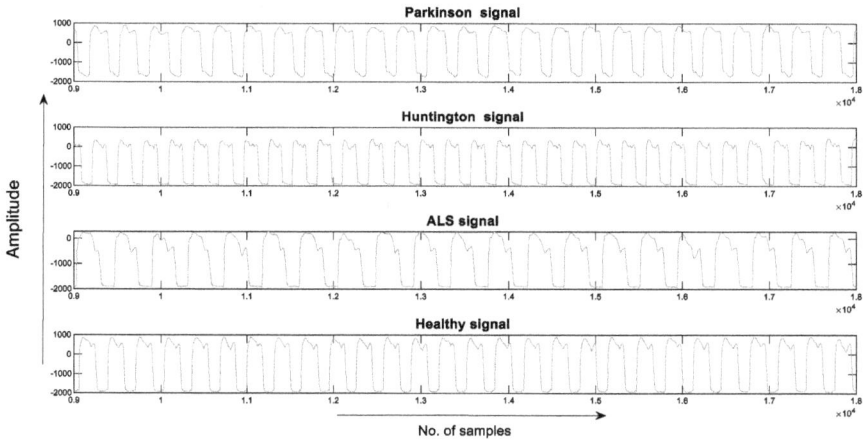

Figure 8.2 Gait time series representation.

8.2.2 Discrete wavelet transforms

Wavelet transforms are extensively used in many disciplines to solve a variety of actual issues. A signal's Fourier transform includes only the spectral spectrum of the gait over the estimation process and thus privations any time domain location information. In order to obtain localization information in time, [29] the time window must be short, compromising localization in frequency. On the contrary, achieving frequency localization necessitates a large time analysis window, which compromises time localization. Therein lays the dilemma, from time to time denoted as the "uncertainty principle". The short-time Fourier transform (STFT) is a kind of negotiation of time and frequency-based perspectives of a signal, and it consists of both time and frequency information. The size of the prediction interval determines the frequency resolution of STFT. This sampling rate is constant across the entire frequency range.

DWT is a characterization of the signal, which is an infinite set of a wavelet on a orthonormal basis. DWT is able to decompose non-linear features of the signal.

In DWT, the dilation and translation measures are discretized, as shown in equation (8.1)

$$\begin{cases} a1 = a1_o^j \\ b = kb_0 \, a1_o^j \end{cases} \quad a1_0 < 1, \quad b_0 \neq 0, \quad j \in Z, \quad k \in Z \tag{8.1}$$

The base wavelet has discretized measures, which can be written as,

$$\Psi_{j,k}(t) = \left(\frac{1}{\sqrt{a_0^j}} * \Psi\left(\frac{t - ka1_0^j b_0}{a1_0^j}\right)\right) \tag{8.2}$$

Therefore, DWT of the signal can be denoted as,

$$D(j, k) = \int_{-\infty}^{+\infty} u1(t)\Psi_{j,k}(t)dt \tag{8.3}$$

The signal u1(t) can be duplicated using the inverse DWT, as given below:

$$u1(t) = \frac{1}{a1} \sum_{j=-\infty}^{J} \sum_{k=-\infty}^{\infty} wt(j, k)\Psi_{j,k}(t)a1 \in R^+ \tag{8.4}$$

The signal is instantaneously carried through LP and HP filters during the initial step of the DWT, also with cut-off frequency being one-fourth of the sampling frequency [30]. The low and high pass filter outcomes are mentioned as the approximation (A) and detail (D) coefficients of the first level, respectively.

According to the Nyquist rule, output signals with half the frequency bandwidth of the input signal can be down sampled by two. To obtain the second-level coefficients, repeat the steps used for the first-level approximation and the detail coefficients. The frequency resolution is doubled through filtering at each step of this decomposition process, and the time resolution is halved through down sampling. Figure 8.3 depicts a signal's nth-level wavelet decomposition.

8.2.3 Feature extraction

Feature extraction is a critical process in the medical field for event categorization of signals and images for diagnostic techniques [31]. In this

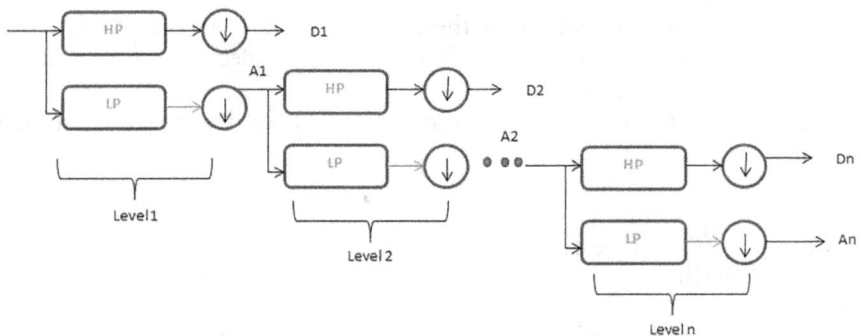

Figure 8.3 Nth level of wavelet decomposition of the gait time series.

current work, entropy and statistical-based feature extraction has been performed for the classification of gait signals for the identification of NDDs. Nine statistical measures – minimum, maximum, mean, standard deviation, energy, normalized energy, kurtosis, skewness, and normalized standard deviation features – have been computed from the wavelet decomposed coefficients. Five entropy measures – log energy entropy, sample entropy, approximate entropy, permutation entropy, and fuzzy entropy features – have been computed for the classification of gait signals. The level of involvement in non-stationary signals is calculated using log energy entropy [32]. That can be mathematically illustrated as

$$\text{Log energy entropy} = \sum_{i=1}^{N} \log(y1_i^2) \tag{8.5}$$

Approximate Entropy assesses the randomness and uniformity of a gait signal across multiple dimensions [33]. It is a scale invariant that describes the similarity of the samples. Sample Entropy is an updated form of Approximate Entropy that is beneficial for complexity reduction in gait signals [34]. It is used to compute the uniformity in the signal more efficiently than Approximate Entropy. It is independent of the length of the time cycles. Permutation Entropy [35] was used to approximate time series complexity by comparing neighboring time series signals. It is a non-stationary time series method that yields efficient, reliable, and quick results. The relationship between the distinguishable number of equally spaced values for a past and present esteem is labelled by mapping the continuous time series record onto a symbolic sequence. Fuzzy Entropy [36] signifies the information of consistency, and it is implemented from the fuzzy sets, whose measure on the gait recordings has a specific function, which is known as membership function with a value remains from 0 to 1. Fuzzy Entropy is a continuous function that is used to estimate resemblance in the time sequence of the signal.

8.2.4 Artificial neural network classifier

Classification is an important area of research in big data, and neural networks are one of the most widely used classification techniques. Three parameters that can make a significant difference are typically used to challenge NDD classification using gait features: 1) a limited number of clinical samples; 2) a large number of noise level or redundant features; and 3) the need to meet real-time deadlines [37]. ANN is a sophisticated adaptive system that could also change its inner core in responding to input. It's done by altering the weight of the correlation. Each link has a specific amount of weight. Weights handle the connection between two neurons, which can be adapted to improve system accurateness.

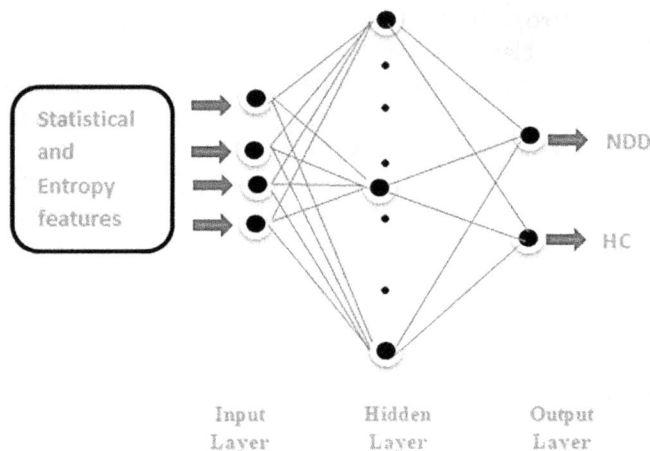

Figure 8.4 Artificial neural network classifier.

The signal between two nerve cells is governed by a weight, which is a scalar. An ANN connects the output layer, input layer, and one or more hidden layers (middle layer) with [38] a larger number of processing nodes or artificial neurons. The function of ANN classifier is given in Figure 8.4.

In the current study, scaled conjugate gradient back propagation is utilized for training the samples. The classification performance of the ANN classifier computed by the following parameters.

$$\text{Classification Accuracy} = \frac{True\,(+) + True\,(-)}{True\,(+) + True\,(-) + False\,(+) + False\,(-)} * 100 \quad (8.6)$$

$$\text{Sensitivity} = \frac{True\,(+)}{True\,(+) + False\,(-)} * 100 \quad (8.7)$$

$$\text{Specificity} = \frac{True\,(-)}{True\,(-) + False\,(+)} * 100 \quad (8.8)$$

$$\text{Positive predictive value}(PPV) = \frac{True\,(+)}{True\,(+) + False\,(+)} * 100 \quad (8.9)$$

$$\text{Negative Predictive Value}(NPV) = \frac{True\,(-)}{True\,(-) + False\,(-)} * 100 \quad (8.10)$$

Where, True (+) indicates True positive, True(-) refers to True negative, False (+) corresponds to False positive and False(-) implies False negative.

8.3 RESULTS AND DISCUSSION

To investigate the proposed technique, gait signals from pressure sensors from the gait dynamics for NDDs dataset are used. The non-linear characteristics and statistical information of gait signals are investigated in the presented work by using wavelet analysis based on time-frequency computation on gait signals. It captures the effective non-linear properties and statistical features from the wavelet decomposed coefficients of a given signal where the optimum statistical-based features such as minimum, maximum, mean, standard deviation, normalized standard deviation, kurtosis, skewness energy, and normalized energy were captured from the left and right feet of a healthy control subject, an ALS patient, an HD patient, and a PD patient, respectively.

Table 8.1 shows the classification performance of the task PD vs. HC, individual statistical measures, and in Table 8.2, entropy measures are considered. When we consider all the features, minimum feature achieved

Table 8.1 Classification results achieved using statistical features for the classification of PD vs. HC

Features	Accuracy (%)	Sensitivity (%)	Specificity (%)	PPV (%)	NPV (%)
Minimum	91.94	87.50	96.67	95.57	89.27
Maximum	61.61	64.38	58.67	62.19	63.13
Mean	91.94	87.50	96.67	95.57	89.27
Standard deviation	64.84	65.00	64.67	69.09	65.70
Normalized standard deviation	75.48	70.63	80.67	79.09	72.52
Kurtosis	76.77	78.75	74.67	77.10	77.03
Skewness	73.87	77.50	70.00	73.49	74.64
Energy	67.42	77.50	56.67	66.29	71.07
Normalized energy	62.26	72.50	51.33	62.16	66.85
All features	99.38	99.23	99.41	99.29	98.65

Table 8.2 Classification results achieved using entropy features for the classification of PD vs. HC

Features	Accuracy (%)	Sensitivity (%)	Specificity (%)	PPV(%)	NPV(%)
Log- Energy	85.36	87.03	71.26	84.73	84.37
Approximate	98.13	96.75	94.88	97.48	97.43
Sample	91.94	87.50	96.67	95.57	89.27
Permutation	85.16	86.25	84.00	85.36	87.03
fuzzy	97.42	96.88	98.00	98.13	96.75
All features	99.78	99.46	99.70	99.75	99.72

Table 8.3 Classification results achieved using statistical features for the classification of HD vs. HC

Features	Accuracy (%)	Sensitivity (%)	Specificity (%)	PPV(%)	NPV(%)
Minimum	58.61	26.25	84.50	57.17	58.94
Maximum	61.67	39.38	79.50	60.03	62.34
Mean	63.61	42.50	80.50	64.63	63.57
Standard deviation	63.33	56.25	69.00	60.37	66.02
Normalized standard deviation	54.72	33.75	71.50	50.51	57.17
Kurtosis	57.78	40.00	72.00	52.97	60.46
Skewness	70.56	69.38	71.50	67.00	74.25
Energy	70.28	60.63	78.00	69.10	71.32
Normalized energy	71.94	79.38	66.00	65.11	80.34
All features	97.17	96.25	98.61	98	98.94

Table 8.4 Classification results achieved using entropy features for the classification of HD vs. HC

Features	Accuracy (%)	Sensitivity (%)	Specificity (%)	PPV(%)	NPV(%)
Log- Energy	72.34	90.53	96.84	94.93	81.67
Approximate	83.57	85.44	50.94	58.28	63.61
Sample	66.02	25.81	57.89	61.99	63.33
Permutation	57.17	76.33	39.72	78.53	54.72
fuzzy	95.46	92.68	94.84	92.70	97.78
All features	99.98	99.22	99.72	99.73	99.71

the maximum classification performance by having accuracy of 91.94%, sensitivity of 87.50%, specificity of 96.67% and PPV of 95.57% with NPV of 89.27%. When all the 9 statistical features fed in the ANN classifier, that achieved the accuracy of 99.38%.

All entropies considered for the classification attained the classification accuracy of 99.78%. Tables 8.3 and 8.4 list the classification performance of HD vs. HC.

Normalized energy from the statistical features and fuzzy entropy from the entropy features attained the maximum classification performance with respect to all the classification parameters considered. Consideration of all statistical and entropy features attained the accuracy of 97.17% and 99.98%, respectively.

Tables 8.5 and 8.6 show the classification performance of the ALS vs. HC, thereby concluding that the proposed method outperforms by using the normalized energy feature from the statistical and fuzzy & approximate entropy measure attained the maximum accuracy of 88.28%, 86.85%, and 84.86%, respectively.

Table 8.5 Classification results achieved using statistical features for the classification of ALS vs. HC

Features	Accuracy (%)	Sensitivity (%)	Specificity (%)	PPV(%)	NPV(%)
Minimum	75.86	86.25	63.08	75.57	76.72
Maximum	85.17	88.13	81.54	85.54	84.86
Mean	65.52	67.50	63.08	69.65	60.81
Standard deviation	84.48	91.88	75.38	82.16	88.45
Normalized standard deviation	76.21	93.75	54.62	71.97	86.85
Kurtosis	72.41	87.50	53.85	70.00	77.78
Skewness	64.83	74.38	53.08	60.16	67.07
Energy	78.28	79.38	76.92	82.30	73.03
Normalized energy	88.28	91.25	84.62	87.93	88.95
All features	98.97	93.75	93.08	97.31	91.50

Table 8.6 Classification results achieved using entropy features for the classification of ALS vs. HC

Features	Accuracy (%)	Sensitivity (%)	Specificity (%)	PPV(%)	NPV(%)
Log- Energy	76.72	50.74	80.24	72.72	75.86
Approximate	84.86	70.03	86.78	84.73	85.17
Sample	60.81	30.52	68.47	65.08	65.52
Permutation	88.45	68.91	86.72	83.19	84.48
fuzzy	86.85	53.31	81.40	71.30	76.21
All features	97.78	94.45	97.78	98.64	92.41

From Tables 8.7 and 8.8 the classification performance of NDD vs. HC is considered with the individual features from both, such as statistical and entropy attained, showing the better classification results. As shown in Figure 8.5, the proposed method outperforms in all the classification tasks.

The three patient groups with NDDs are at various stages in the current database. They are assessed by the severity or length of the illness. This is the Hohn and Yahr score for PD patients. This is the total functional storage assessment for subjects with HD.

The number here represents the number of months since the subject's diagnosis, demonstrated in the experiments that patients with NDDs, even those in the early stages, can be categorized. This validated the proposed method's efficacy in early detection. Nonetheless, more clinical research and data analysis with larger sample sizes are needed to verify the method's efficacy and appropriateness in the early stages.

Table 8.7 Classification results achieved using statistical features for the classification of NDD vs. HC

Features	Accuracy (%)	Sensitivity (%)	Specificity (%)	PPV(%)	NPV(%)
Minimum	58.94	63.13	35.79	46.19	58.61
Maximum	62.34	70.53	46.84	54.93	61.67
Mean	63.57	75.44	50.94	58.28	63.61
Standard deviation	66.02	65.81	57.89	61.99	63.33
Normalized standard deviation	57.17	76.33	39.72	48.53	54.72
Kurtosis	60.46	76.68	44.84	52.70	57.78
Skewness	74.25	41.06	67.91	70.21	70.56
Energy	71.32	39.50	64.38	68.61	70.28
Normalized energy	80.34	45.41	71.46	72.29	71.94
All features	92.39	92.22	92.97	93.37	94.70

Table 8.8 Classification results achieved using entropy features for the classification of NDD vs. HC

Features	Accuracy (%)	Sensitivity (%)	Specificity (%)	PPV(%)	NPV(%)
Log- Energy	96.67	95.57	89.27	84.50	91.04
Approximate	64.67	69.09	65.70	31.18	62.58
Sample	80.67	79.09	72.52	51.45	74.26
Permutation	74.67	77.10	77.03	53.77	77.67
fuzzy	93.00	93.49	94.64	96.81	94.35
All features	97.00	95.36	97.03	91.26	94.73

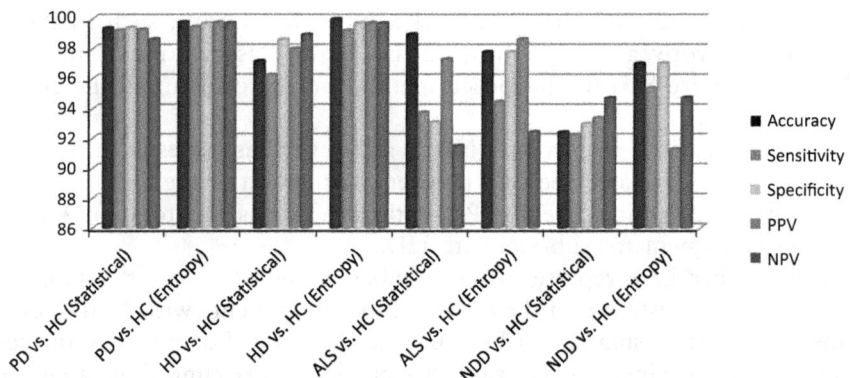

Figure 8.5 The summary of classification performance considering all features for the classification.

8.4 CONCLUSION

The central nervous system is crucial in the regulation of human gait. PD is a common NDD that can cause neurophysiologic changes in the central nervous system, altering the duration of the gait cycle (stride interval). The time-frequency based wavelet analysis performed on the gait signals by applying discrete wavelet transform with the decomposed detail and approximate coefficients. The nine statistical and five entropy measures from the decomposed coefficients computed for the automated detection of NDDs. These optimum features are classified using ANN classifier with the 10-fold cross validation with the better training and testing. The classification performances of the four different tasks are validated by incorporating every single statistical and entropy feature, also computed by considering all the entropy measures and statistical features.

REFERENCES

[1] Hausdorff, Jeffrey M., Apinya Lertratanakul, Merit E. Cudkowicz, Amie L. Peterson, David Kaliton, and Ary L. Goldberger. "Dynamic markers of altered gait rhythm in amyotrophic lateral sclerosis." *Journal of Applied Physiology* (2000).
[2] Viteckova, Slavka, Patrik Kutilek, Zdenek Svoboda, Radim Krupicka, Jan Kauler, and Zoltan Szabo. "Gait symmetry measures: A review of current and prospective methods." *Biomedical Signal Processing and Control* 42 (2018): 89–100.
[3] Moon, Yaejin, JongHun Sung, Ruopeng An, Manuel E. Hernandez, and Jacob J. Sosnoff. "Gait variability in people with neurological disorders: A systematic review and meta-analysis." *Human movement science* 47 (2016): 197–208.
[4] Aziz, Wajid, and Muhammad Arif. "Complexity analysis of stride interval time series by threshold dependent symbolic entropy." *European Journal of Applied Physiology* 98 (2006): 30–40.
[5] Joshi, Deepak, Aayushi Khajuria, and Pradeep Joshi. "An automatic non-invasive method for Parkinson's disease classification." *Computer Methods and Programs in Biomedicine* 145 (2017): 135–145.
[6] Chen, J. Jean. "Functional MRI of brain physiology in aging and neurodegenerative diseases." *Neuroimage* 187 (2019): 209–225.
[7] Blijham, Paul J., H. Jurgen Schelhaas, Henk J. Ter Laak, Baziel G.M. van Engelen, and Machiel J. Zwarts. "Early diagnosis of ALS: The search for signs of denervation in clinically normal muscles." *Journal of the Neurological Sciences* 263, no. 1-2 (2007): 154–157.
[8] Baratin, E., L. Sugavaneswaran, K. Umapathy, C. Ioana, and S. Krishnan. "Wavelet-based characterization of gait signal for neurological abnormalities." *Gait & Posture* 41, no. 2 (2015): 634–639.
[9] Bilgin, Suleyman. "The impact of feature extraction for the classification of amyotrophic lateral sclerosis among neurodegenerative diseases and

healthy subjects." *Biomedical Signal Processing and Control* 31 (2017): 288–294.

[10] Hausdorff, Jeffrey M., Susan L. Mitchell, Renee Firtion, Chung-Kang Peng, Merit E. Cudkowicz, Jeanne Y. Wei, and Ary L. Goldberger. "Altered fractal dynamics of gait: reduced stride-interval correlations with aging and Huntington's disease." *Journal of Applied Physiology* 82, no. 1 (1997): 262–269.

[11] Pham, Tuan D. "Texture classification and visualization of time series of gait dynamics in patients with neuro-degenerative diseases." *IEEE Transactions on Neural Systems and Rehabilitation Engineering* 26, no. 1 (2017): 188–196.

[12] Prabhu, Pooja, A. Kotegar Karunakar, H. Anitha, and N. Pradhan. "Classification of gait signals into different neurodegenerative diseases using statistical analysis and recurrence quantification analysis." *Pattern Recognition Letters* 139 (2020): 10–16.

[13] Morris, Meg E., Jennifer McGinley, Frances Huxham, Janice Collier, and Robert Iansek. "Constraints on the kinetic, kinematic and spatiotemporal parameters of gait in Parkinson's disease." *Human Movement Science* 18, no. 2-3 (1999): 461–483.

[14] Ghaderyan, Peyvand, and Seyede Marziyeh Ghoreshi Beyrami. "Neuro-degenerative diseases detection using distance metrics and sparse coding: A new perspective on gait symmetric features." *Computers in Biology and Medicine* 120 (2020): 103736.

[15] Dutta, Saibal, Amitava Chatterjee, and Sugata Munshi. "An automated hierarchical gait pattern identification tool employing cross-correlation-based feature extraction and recurrent neural network based classification." *Expert Systems* 26, no. 2 (2009): 202–217.

[16] Hausdorff, Jeffrey M. "Gait variability: methods, modeling and meaning." *Journal of Neuroengineering and Rehabilitation* 2, no. 1 (2005): 1–9.

[17] Wu, Yunfeng, and Sridhar Krishnan. "Computer-aided analysis of gait rhythm fluctuations in amyotrophic lateral sclerosis." *Medical & Biological Engineering & Computing* 47 (2009): 1165–1171.

[18] Daliri, Mohammad Reza. "Automatic diagnosis of neuro-degenerative diseases using gait dynamics." *Measurement* 45, no. 7 (2012): 1729–1734.

[19] Wu, Yunfeng, and Sridhar Krishnan. "Statistical analysis of gait rhythm in patients with Parkinson's disease." *IEEE Transactions on Neural Systems and Rehabilitation Engineering* 18, no. 2 (2009): 150–158.

[20] Wu, Yunfeng, and Lei Shi. "Analysis of altered gait cycle duration in amyotrophic lateral sclerosis based on nonparametric probability density function estimation." *Medical Engineering & Physics* 33, no. 3 (2011): 347–355.

[21] Moustakidis, S.P., J.B. Theocharis, and G. Giakas. "Feature selection based on a fuzzy complementary criterion: application to gait recognition using ground reaction forces." *Computer Methods in Biomechanics and Biomedical Engineering* 15, no. 6 (2012): 627–644.

[22] Hu, Haifeng. "Multiview gait recognition based on patch distribution features and uncorrelated multilinear sparse local discriminant canonical correlation analysis." *IEEE Transactions on Circuits and Systems for Video Technology* 24, no. 4 (2013): 617–630.

[23] Zeng, Wei, Cong Wang, and Feifei Yang. "Silhouette-based gait recognition via deterministic learning." *Pattern Recognition* 47, no. 11 (2014): 3568–3584.

[24] Wagenaar, Robert C., and Richard E.A. van Emmerik. "Dynamics of movement disorders." *Human Movement Science* 15, no. 2 (1996): 161–175.

[25] Hausdorff, Jeffrey M., Merit E. Cudkowicz, Renée Firtion, Jeanne Y. Wei, and Ary L. Goldberger. "Gait variability and basal ganglia disorders: stride-to-stride variations of gait cycle timing in Parkinson's disease and Huntington's disease." *Movement Disorders* 13, no. 3 (1998): 428–437.

[26] Xia, Yi, Qingwei Gao, and Qiang Ye. "Classification of gait rhythm signals between patients with neuro-degenerative diseases and normal subjects: Experiments with statistical features and different classification models." *Biomedical Signal Processing and Control* 18 (2015): 254–262.

[27] Zeng, Wei, and Cong Wang. "Classification of neurodegenerative diseases using gait dynamics via deterministic learning." *Information Sciences* 317 (2015): 246–258.

[28] Hausdorff, Jeffrey M., Zvi Ladin, and Jeanne Y. Wei. "Footswitch system for measurement of the temporal parameters of gait." *Journal of Biomechanics* 28, no. 3 (1995): 347–351.

[29] Ocak, Hasan. "Automatic detection of epileptic seizures in EEG using discrete wavelet transform and approximate entropy." *Expert Systems with Applications* 36, no. 2 (2009): 2027–2036.

[30] Prasanna, J., M.S.P. Subathra, Mazin Abed Mohammed, Robertas Damaševičius, Nanjappan Jothiraj Sairamya, and S. Thomas George. "Automated epileptic seizure detection in pediatric subjects of CHB-MIT EEG database—a survey." *Journal of Personalized Medicine* 11, no. 10 (2021): 1028.

[31] Jothiraj, Sairamya Nanjappan, Thomas George Selvaraj, Balakrishnan Ramasamy, Narain Ponraj Deivendran, and Subathra M.S.P. "Classification of EEG signals for detection of epileptic seizure activities based on feature extraction from brain maps using image processing algorithms." *IET Image Processing* 12, no. 12 (2018): 2153–2162.

[32] Subathra, M.S.P., Mazin Abed Mohammed, Mashael S. Maashi, Begonya Garcia-Zapirain, N.J. Sairamya, and S. Thomas George. "Detection of focal and non-focal electroencephalogram signals using fast Walsh-Hadamard transform and artificial neural network." *Sensors* 20, no. 17 (2020): 4952.

[33] Acharya, U. Rajendra, Hamido Fujita, Vidya K. Sudarshan, Shreya Bhat, and Joel E.W. Koh. "Application of entropies for automated diagnosis of epilepsy using EEG signals: A review." *Knowledge-based Systems* 88 (2015): 85–96.

[34] Cuesta–Frau, David, Pau Miró–Martínez, Jorge Jordán Núñez, Sandra Oltra–Crespo, and Antonio Molina Picó. "Noisy EEG signals classification based on entropy metrics. Performance assessment using first and second generation statistics." *Computers in Biology and Medicine* 87 (2017): 141–151.

[35] Cao, Yinhe, Wen-wen Tung, J.B. Gao, Vladimir A. Protopopescu, and Lee M. Hively. "Detecting dynamical changes in time series using the permutation entropy." *Physical Review E* 70, no. 4 (2004): 046217.

[36] Chen, Weiting, Zhizhong Wang, Hongbo Xie, and Wangxin Yu. "Characterization of surface EMG signal based on fuzzy entropy." *IEEE Transactions on Neural Systems and Rehabilitation Engineering* 15, no. 2 (2007): 266–272.

[37] Sharma, Bhavna, and K.J.I.J.C.E. Venugopalan. "Comparison of neural network training functions for hematoma classification in brain CT images." *IOSR Journal of Computer Engineering* 16, no. 1 (2014): 31–35.

[38] Subasi, Abdulhamit, M. Kemal Kiymik, Ahmet Alkan, and Etem Koklukaya. "Neural network classification of EEG signals by using AR with MLE pre-processing for epileptic seizure detection." *Mathematical and Computational Applications* 10, no. 1 (2005): 57–70.

Chapter 9

EEG artifact detection and removal techniques

A brief review

Sandhyalati Behera and Mihir Narayan Mohanty

Department of Electronics and Communication Engineering, ITER,
Sikhsha 'O' Anusandhan (Deemed to be University), Bhubaneswar, India

9.1 INTRODUCTION

By attaching metal electrodes to a patient's scalp, doctors can record the brain's electrical activity in real time during a diagnostic procedure known as an electroencephalogram (EEG). When the nervous system is at rest, the brain neurons are still actively engaged in their normal, unprompted process of exchanging information via the production of electrical currents. EEG is an effective tool for brain imaging due to its low cost, high flexibility, high temporal resolution, non-invasiveness, ease of use, portability, and safety in comparison to other functional neuroimaging techniques such as positron emission tomography (PET), magnetoencephalography (MEG), functional magnetic resonance imaging (fMRI), and transcranial magnetic stimulation (TMS). Brain signals, unlike other bodily signals, exhibit periodic patterns in their electrical changes, which can be used to infer the occurrence of some activity, such as the change from sleep to wakefulness, which is characterized by slow, very low frequency, high-amplitude waves (fast, high-frequency, low-amplitude waves). Furthermore, the impact of anesthesia and epilepsy are investigated on EEGs.

The brain's electrical activity, as measured by an electroencephalograph, is a non-linear and non-stationary signal. We can infer the occurrence of some neuronal activity from the slight shifts in the voltage fluctuations measured by EEG. Consequently, the visual examination of these signals varies with the expertise level. Furthermore, manual review of long EEG recordings takes a significant amount of time, and results may be inaccurate due to artifacts in the signals. Thus, computer-aided technologies can be used to process and analyze these signals, yielding quick and precise results. Particularly in the diagnosis of epilepsy [1,2], major depressive disorder (MDD) [3,4], alcohol use disorder (AUD) [5,6], and dementias [7,8] like Alzheimer's, mild cognitive impairment (MCI), Parkinson's, and dementia with Lewy bodies (DLB), the use of computer-aided technologies with EEG signals has gained widespread popularity. The integration of motor imagery into neuro prostheses [9,10] through EEG has led to exciting new

DOI: 10.1201/9781003398066-9

possibilities. In addition, the use of physiological data like EEG is yielding significant results in other research domains, such as identity authentication [11,12], sleep stage classification, emotion recognition, eye state detection, and drowsiness monitoring.

However, artifacts [13,14] are an inherent problem with EEG recordings. As the original EEG activity is distorted, analyzing it becomes more challenging due to artifacts, which are unwanted signals that originate from sources other than neurons. Eye movements, eye blinks, muscle activity, and cardiac activity are common sources of contamination in the EEG, despite the fact that only neuronal activity should be present [15,16]. Because artifact contamination modifies the genuine EEG signal, it also impacts the outcomes of the applied analysis. For instance, it is well established that artifacts lower classification accuracy and controllability. Therefore, it is essential for clinical or applied research to address these artifacts before analyzing EEG signals. This requires a technique that not only effectively eliminates artifacts but also protects the authentic, undistorted neuronal activity present in EEG signals.

To achieve this goal, a number of techniques, both manual and automated, have been created and put into use. This paper provides a comprehensive overview of the difficulties encountered by artifact removal algorithms when processing EEG data. This paper is the first of its kind to elaborate on the algorithm-specific and general difficulties of EEG artifact removal algorithms, to the best of our knowledge. Recent reviews [17–21] have only focused on the specifics of implementing various methods. Artifact removal algorithms are discussed in Jiang et al. [22] and Islam et al. [19,23] with respect to only four algorithm-specific challenges: additional reference channel, automatic, online, and applicability to a single EEG channel. Similarly, Mannan et al. [24] discussed various difficulties associated with artifact removal algorithms in the discussion section. The challenges of creating and evaluating EEG artifact removal algorithms are becoming increasingly apparent, and researchers are beginning to realize they need more information about them. The purpose of this paper is to highlight the challenges inherent in existing methods for cleaning EEG data of artifacts and to propose alternatives for addressing those challenges. The writing of this paper is broken up into a few different parts. In Section 9.2, we delve deeper into the common artifacts found in EEG data. Artifact removal and storage methods are briefly covered in Section 9.3. Sections 9.4 and 9.5 discuss the specific and general challenges of EEG artifact removal algorithms, respectively. In Section 9.6, we offer some suggestions for dealing with these problems.

9.2 DIFFERENT TYPES OF EEG ARTIFACTS

When designing or choosing algorithms to remove artifacts from EEG signals, it is important to have a foundational understanding of the various types of artifacts that can be present. Specifically, artifacts are defined as "unwanted signals that may introduce changes in the measurements and affect the signal of interest" [25]. The subject's physiological activities and movement, as well as environmental interferences (e.g., the movement of electrodes and cables) can introduce both frequency and time domain artifacts into an EEG recording [26]. Cables, which act as antennas when in use, can be shielded and grounded to reduce external artifacts. They can also be isolated and relocated away from recording locations. Contrarily, internal or physiological artifacts present difficulties for scientists. Artifacts from the eyes (called electrooculograms, or EOGs), artifacts from the muscles (EMG), cardiac artifacts (ECG artifacts), and motion artifacts are the most noticeable in a standard EEG recording.

9.2.1 Ocular (EOG) artifact

The most prevalent artifact encountered when measuring EEG signals is the ocular artifact. Because of the potential disparity between the cornea and the retina, an artifact is produced. Blinking and other eye movements alter the potentials in the eye. In particular, ocular conductance artifacts are caused by changes in contact between the cornea and eyelid during blinking [27]. Artifacts caused by eye movement occur when the retina and cornea dipole are no longer aligned. Because of the volume conduction effect, the electrodes were able to pick up both the ocular artifact and the EEG activity occurring on the surface of the subject's head. The electrooculogram is a tool for recording these eye waves (EOG). While the frequencies of EOG and EEG are similar, the amplitude of EOG is typically much higher [27–29]. It's important that both EEG and EOG have the potential to contaminate the other's data. As a result, we will encounter removal error when eliminating EOG artifacts due to bidirectional interference [30].

9.2.2 Muscular artifact

Muscle or electromyographic (EMG) artifact is another source of noise in EEG signals. The action of the frontalis and temporalis muscles is to blame for these. Non-stationary characteristics [31] are inherent to EMG because it is produced by a collection of muscles that are both physically and functionally separated from one another. Muscle artifact waveforms and amplitudes vary by muscle type, degree of contraction, and participant sex [32,33]. It is generally agreed that the frequency range of EMG activity is much larger than that of traditional EEG rhythms. According to [25], the power density of contracting striated muscle is highest at the lower end of the frequency spectrum, which spans from 20 to 300 Hz. Muscle artifacts are most likely to occur in the gamma band, which spans 30 to 40 Hz, given that the peak frequencies of the temporal, masseter, and frontalis muscles are 40 to 80 Hz, 50 to 60 Hz, and 30 to 40 Hz, respectively [31,34].

The beta band overlaps with the frequencies reported in [31] for the frontal muscles, which are around 20–30 Hz. Furthermore, 2 Hz has been reported as the lowest frequency of muscle activity. Muscle artifacts also disrupt the delta, theta, and alpha frequency bands. Therefore, using only the most fundamental spectral signatures, it is challenging to tell EMG apart from EEG.

9.2.3 Cardiac artifact

The heart potential over the scalp causes artifacts in an EEG called electro-cardiogram (ECG). This type of artifact is seen more in cases of a short person with a wide neck. Placing electrodes on or near a blood vessel [31], which is expanding and contracting in response to the beating of the heart, introduces cardiac artifacts. Similar to cardiac artifact, pulse artifacts, with a frequency of about 1.2 Hz, can appear within the EEG as a similar waveform and are therefore challenging to remove [29,35]. In contrast to pulse artifacts, the cardiac artifact can be measured with a characteristic regular pattern [36] and recorded independently of cerebral activity, suggesting that removing such artifacts may be as simple as using a reference waveform. Because of cardiac activity, the EEG signals are distorted in a static magnetic field, causing BCG artifacts. Electrode motion is typically brought on by cardiac-related activities [37] and in 3.0T, the magnitude of the BCG artifacts can reach 400 V (roughly 6–8 times that of EEG) [38].

9.2.4 Motion artifact

Removal of motion artifact (MA) recently gained interest among researchers in the past two years. When the subject's head rotates in relation to the magnetic field, MAs are introduced into the EEG data that were collected at the same time as the fMRI [39]. Because they have the same frequency spectrum as the EEG signal (up to 50 Hz) but a larger amplitude than the brain signals, MAs have a relatively greater impact on ambulatory EEG [40]. These MAs shared a characteristically low frequency with the EEG's Delta rhythm. It's also closely linked to how the subjects and equipment in an experiment are moving around. Since MAs introduce significant variation into the EEG data and distort the original shape of the signal, they must be eliminated [41]. Due to this, incorrect diagnoses of diseases, false alarms, and other problems can result from analyzing EEG data. Voltage changes occur when the electrodes shift from their normal position or the cables sway while the patient is in motion. Various factors, such as ground reaction forces, cyclic motion, head movements, etc., can cause MAs. Therefore, raw EEG data, which can range from high to low power spectral density, is severely corrupted by MAs if recordings are made while the subject is engaged in everyday activities.

9.3 ARTIFACT REMOVAL TECHNIQUES

Artifacts in EEG data are typically eliminated using EEG artifact removal strategies. Due to the effectiveness of artifact removal techniques for EEG data, the techniques can now be used in a wide variety of clinical and industrial settings. There have been a number of obstacles in the way of EEG artifact removal techniques. These difficulties may arise from the nonlinearities of the unwanted signal being added to the EEG signal or from the complexity of the methods themselves. The "nonlinear" nature of the artifacts, for instance, makes it challenging to extract only the artifacts without also losing actual neuronal data. In this literature, various artifact removal algorithms are discussed in the following subsections.

9.3.1 Regression technique

Since the middle of the 1990s, a number of methods have been proposed to remove different artifacts from EEG data, but the linear regression method has gained popularity due to its ease of use. Each EEG channel is thought to record a linear sum of natural brain activity and artifact signals. The signals from these artifacts can be retrieved with the help of reference channels or artifact templates. If the contaminated EEG signal is subtracted from a regressed portion of the reference signal, the artifacts will be eliminated. As such, the goal of regression algorithms is to provide an accurate estimate of the optimal value that characterizes the extent to which the reference channel (i.e., the artifact) affects the EEG signals [42]. In order to recover EEG data from ocular noise, linear regression has been used extensively.

A common method for detecting artifactual samples and subsequently removing them from the model is regression analysis [28,43], which employs a multi-modal linear model between the observed and a reference signal. Ocular artifacts can be eliminated with the help of time-domain regression [44]. Frequency domain regression [45] was also introduced, but neither of these methods could deal with the inherent problem of bidirectional contamination in EEG data [28]. This issue can be fixed, however, by employing the filtering method first, followed by the regression [46–49]. This strategy is justified on the grounds that filtering out the high-frequency components of recorded EOG will drastically cut down on the bidirectional contamination effect, as most of that content belongs to the neuronal activity [48]. While methods based on regression and blind source separation (BSS) [43] are able to eliminate the issue of bidirectional contamination, they are severely constrained.

The regression technique was used to remove the ocular and cardiac artifacts. But when there is no reference channel, such regression analysis frequently fails. Regression methods have been replaced by more advanced methodologies [25,50,51] due to their reliance on a reference channel, which restricts their use to EOG and ECG primarily. In addition, the method assumes incorrectly that the neuronal activity in EEG and EOG signals are uncorrelated, while attempting to remove artifacts using EOG signals as reference [25,51]. Thus, the shared neuronal activity between EEG and EOG can be removed from EEG signals using regression analysis. There is currently no agreement in the scientific literature regarding the best approach for low-pass filtering EOG signals. On the other hand, some authors contend that neural activity pollutes every frequency range [52]. However, regression methods continue to serve as the benchmark against which the efficacy of all other newly developed methods must be judged.

9.3.2 Filtering technique

Filtering technique is one of the popular artifact removal methods from the EEG signal. Generally, two types of filtering technique are used in this case. Classical filters can only be applied when the frequency bands of the noise and the signal of interest do not substantially overlap. However, the frequency bands of muscle artifact and brain signals often do overlap, which is the main reason for failure of classical filtering in muscle artifact removal. To overcome this limitation, there are a number of new filtering techniques applied for artifact elimination, such as adaptive filtering [53] and Kalman filtering [54]. In what follows, we briefly introduce the main two filtering techniques.

9.3.2.1 Adaptive filtering

The adaptive filtering presupposes that the target signal and the artifact are independent of one another. Based on a reference signal, the filter taps a delay line to produce a signal that is correlated with the true artifact signal. The estimated artifact is then subtracted from the recorded signal, and the resulting residual is a proxy for the true signal [55]. When an artifact is detected in the output and it is found to be correlated with the reference, the filter coefficients will continue to adjust until the artifact is eliminated. Successful application of adaptive filtering relies heavily on selecting an appropriate reference signal [50] and also on the choice of an appropriate algorithm to update the filter weights. The algorithms used to update the weights are least mean squares (LMS) and variants of LMS, recursive least squares (RLS) and variants of RLS, and state space RLS.

In the case of adaptive filtering, to update the filter weights, reference signal is used as one of the inputs. Based on reference input, the adaptive filter produces a signal, which is the correlated version of reference input. The output of the filter is subtracted from the contaminated EEG signal to produce an appropriate signal of interest. The electrooculogram (EOG) signal is used as a reference to filter out ocular artifacts in EEG data [56], and/or reference signals can be measured using an electrocardiogram (ECG) for filtering out cardiac artifacts [57]. The artifacts in EEG signals can be estimated most accurately with the help of an optimization algorithm. The most widely used adaptive algorithm for modifying a weight vector is the LMS algorithm [58]. Recursive least squares (RLS)-based adaptive filtering is another popular algorithm [56,59].

The second-order nature of the RLS algorithm results in very quick convergence. There are a number of papers that detail the benefits of both families of algorithms, including least mean absolute value and sign (SIGN) [60]. The RLS family of algorithms consistently outperforms the competition in terms of accuracy, but at the expense of a higher computational cost [60,61]. Adaptive filters have a few advantages, such as online implementation, no preprocessing/calibration, and ease of use, but one of their limitations is that they need a reference signal using additional sensors.

Other filters, such as Kalman, Wiener, and Bayes filters, can also be used for artifact removal; however, these methods have not been extensively explored in the literature of EEG artifact removal [25,62–65]. When unwanted artifacts are present in the measured signals, Wiener filtering can be used as a parametric technique to eliminate them [66]. Unlike other algorithms, the Wiener algorithm can function without any kind of external reference signal because it is based on statistical principles. It is assumed that the signal and the (additive) artifact are both stationary linear stochastic processes with well-characterized spectra or well-established patterns of autocorrelation and cross-correlation. One disadvantage of the Wiener filter over the adaptive filter is that it requires calibration before use and can't be used in real-time applications. The adaptive filter, on the other hand, does not necessitate any supplementary hardware on the recording device.

Using a predictor-corrector strategy, Bayes filters can be built. The predictor describes the connection between states from one time sample to the next using a time update model. Afterwards, a measurement model is used in the corrector phase to characterize the connection between the external data and the internal state. The Bayes filter technique, like Wiener filtering, can eliminate the embedded artifact without the use of a reference signal. In contrast to the Wiener algorithm, Bayes filtering can function in real time. Several approaches exist for realizing the various approximations of the Bayes filter technique used in modern signal processing. Since the algorithm for the Bayes filter is computationally intractable, the filtering method itself is not implemented. Bayes filter is implemented by Kalman filter [67–70] and Particle filter [71,72].

To estimate a process, the Kalman filter employs feedback control by first making an estimate of the state of the process at a given time and then receiving information about the process in the form of (noisy) measurements [67]. Kalman filters can be used in a variety of ways to remove artifacts [73–75]. The first step is to create a model of both the target signal and the noise. The recorded signal can be understood in terms of the sum of these two model signals, allowing one to infer both the process and measurement models. Kalman filters are a great alternative to adaptive filters if process and measurement models for the necessary system are already available.

The use of the particle filter to eliminate artifacts in physiological signals [76] is an area that has seen surprisingly little research. The particle filter, on the other hand, can remove artifacts in a way that's analogous to the Kalman filter's success. For optimal performance, Kalman filters necessitate sensors with both high accuracy and a high rate of update. When compared to other Bayes' method-based filtering techniques, they are one of the most time- and space-efficient. However, if precise sensors are unavailable or if a comprehensive model of the process and measurements is lacking, particle filtering emerges as the preferred method. Particle filters are a versatile tool with low implementation overhead because they don't need a detailed model of the systems [77].

9.3.3 Decomposition technique

Separating the EEG and artifact sources into distinct components and then eradicating them during reconstruction is another denoising method. Techniques for decomposing EEG data include BSS, wavelet transform (WT), empirical mode decomposition (EMD), and variational mode decomposition (VMD).

9.3.3.1 Techniques of blind source separation (BSS)

BSS is based on a wide variety of unsupervised learning algorithms [78] with the intention of estimating sources (which are not necessarily independent) and mixing system parameters. Because it can decouple the source signals of neuronal activity from the artifacts, it is one of the most well-known and widely used methods for extracting EEG data [25,26,50]. When it comes to blending multiple sources together, BSS shines because it doesn't necessitate any prior knowledge (or, in some cases, uses very limited knowledge). Independent component analysis (ICA), canonical correlation analysis (CCA), morphological component analysis (MCA), and independent vector analysis (IVA) are just some of the algorithms available for BSS.

9.3.3.1.1 Independent component analysis (ICA)

ICA separates multichannel EEG data into its constituent parts called independent components (ICs). ICA is used on the premise that the signals coming from various sources are linearly mixed. Because it avoids the problems that plague parametric methods like adaptive filtering, ICA has recently emerged as a useful tool for cleaning up EEG data. EEG artifact removal studies employ ICA more frequently than any other BSS technique [79–84]. JADE [16], fast ICA, SOBI, InfoMax [85], constrained ICA [86],

AMICA [87], and AMUSE [88] are just a few of the ICA variants proposed by researchers for EEG artifact removal.

The necessity of using non-Gaussian data for ICA is a significant limitation of the method. If the sources are not Gaussian, ICA can be used to estimate them. ICA can only account for a single Gaussian component, which can be estimated as the residual after all other independent components have been extracted. The Gaussian or non-Gaussian nature of a given component is rarely known in advance. Normalizing the data is a standard practice before ICA is calculated. Another drawback of ICA is that it requires a number of channels that is at least as large as the number of sources, so it can't be used with just one or a small number of channels.

9.3.3.1.2 Canonical correlation analysis (CCA)

In order to discover the nature of the connection between two datasets, CCA was developed as a statistical technique. Second-order statistics (SOS) are utilized by the CCA method to generate components that are intrinsically uncorrelated. CCA eliminates the BSS issue by making all sources maximally autocorrelated and uncorrelated with one another [89].

The recorded EEG data and its time-delayed version are used as the first data set and second data set, respectively, for CCA's application to the BSS problem. CCA is a solution to the BSS problem because it ensures that the canonical variates are maximally correlated across both data sets while remaining independently correlated within each [90]. Since muscle artifacts have a lower autocorrelation than other types of artifacts, they can be removed from a reconstructed image by setting the autocorrelation of the final several source components to zero.

9.3.3.1.3 Morphological component analysis (MCA)

MCA functions by breaking down the recorded signal into sub-signals with distinct morphological properties. A collection of waveforms (called atoms) is used to describe the various underlying signals, and these are sparsely represented in an overcomplete dictionary. Consequently, the total signal is simply the product of these atoms multiplied by their respective coefficient vectors in a linear fashion. The MCA algorithm was used to clean up a single-channel EEG recording from artifacts like brow furrowing, jaw clenching, swallowing, and blinking eyes by the authors of [91].

9.3.3.1.4 Principal component analysis (PCA)

Principal component analysis (PCA) is an often-used BSS technique whose algorithm relies on the eigen values of a covariance matrix [22]. The first step in this process is to use orthogonal transformation to transform any correlated variables into independent ones. These sets of independent variables are referred to as principal components (PCs). EEG signal PCs will be implemented via single value decomposition (SVD). According to [92], PCA of variance is used to extract the main components standing for blinks and eye movements. After that, we used an inverse operation to eliminate the associated components and obtain the clean EEG data. It was shown in [93] that PCA is a more computationally efficient alternative to linear regression. On the other hand, it can be challenging to ensure that artifact components are independent of EEG readings. PCA also fails to discriminate interferences when the potential of drifts and EEG data are similar. The subsequent studies, therefore, favor more adaptable approaches like ICA.

9.3.3.2 Wavelet transform (WT)

WT is one of the decomposition techniques that acts as a sub-band filter. Use of this transform eliminates high-frequency signal to restore the low-frequency signal and is applied to remove the artifacts from EEG signal. The artifacts, such as ocular artefacts in EEG [94–96] and EMG artifacts in fNIRS. Decomposing the signal into shifted and scaled versions of the wavelet basis [97] is the first step in the method for eradicating artifacts in EEG, after which the remaining signal components are cleaned up and reconstructed. The time-frequency analysis of EEG signals commonly employs the complex Morlet wavelet as the wavelet basis [98]. Although the discrete wavelet transform (DWT) is a fast and effective method, its major drawback is its time-variance, which is very important for EEG data [99]. This is because DWT uses iterative high and low pass filters to represent signals into its discrete wavelets. Although slower than DWT [100], stationary wavelet transform (SWT) gets around DWT's weakness in translation invariance. The key distinction between the two is that DWT uses a sampling factor of $2^j - 1$ to reconstruct the signal while SWT uses a sampling factor of 2j to down-sample the filter output before reconstructing it. When the measured signal's spectral properties overlap with the spectral properties of the artefacts, wavelet-based methods are unable to remove the artifacts completely [25,26].

9.3.3.3 Empirical mode decomposition (EMD)

As similar to time-frequency WT technique, EMD is another decomposition method for this application. This method is an adaptive and flexible data-driven method that splits a time-domain signal into a collection of intrinsic mode functions (IMFs) [101,102]. Artifacts in EEG data can be removed using EMD alone [103] or in conjunction with other techniques [101,104]. Due to its sensitivity to background noise, EMD has undergone refinements to address the challenges posed by mode mixing. A noise-assisted data analysis method, enhanced EMD (EEMD) is developed with the average number of IMFs from EMD as the optimal IMFs [105,106].

9.3.3.4 Variational mode decomposition

Modification of EMD is termed variational mode decomposition (VMD), where variation of center frequency needed to generate IMFs. It is a relatively new non-recursive signal decomposition method that can be used to infer the structure of a non-stationary signal [107,108]. VMD is a method for breaking down a multi-component EEG signal into smaller, more manageable pieces, or "modes," each of which has its own unique spectral density. Despite being effective at reducing single-channel OAs, most methods for eliminating OAs also distort important clinical aspects of the EEG signal in the process. In [109] the authors propose a reliable framework for the detection and removal of OAs using VMD and turning point count to address these concerns. The VMD procedure consisted of two phases, VMD-I and VMD-II. Rejecting low-frequency baseline components from the raw EEG signal using VMD-I; dividing the processed EEG signal into three modes using VMD-II; and finally, rejecting the mode that contains OAs using a threshold based on the number of turning points in the signal [109].

To process signals in an adaptive and non-recursive way, VMD is a popular choice. To a large extent, the outcome of signal decomposition is controlled by two parameters in VMD. Parameter-related problems are reduced thanks to a revised VMD based on the squirrel search algorithm (SSA). The method is used to fine-tune VMD's critical settings. To filter out eye movements and other visual noise from EEG recordings, the GOSSA-VMD model is developed [110].

Although the amplitude of an eye blink is greater than that of a typical EEG signal, a standard peak-detection algorithm applied to the raw EEG signal would likely generate numerous false positives. Blink regions in IMFs were sought after by researchers using an extension of VMD [107], called multivariate variational mode decomposition (MVMD) [111,112].

9.3.4 Machine learning technique

Algorithms based on machine learning can also be used to clean up EEG data that has been contaminated by artifacts. In order to detect artifacts in EEG data using traditional methods, trained observers are required. Using machine learning based algorithms, the artifactual patterns can be detected and eliminated mechanically. This has the potential to greatly improve the integrity of the original EEG data while simultaneously enhancing artifact removal accuracy. Artificial neural networks (ANN) and support vector machines (SVMs) are the most common machine learning based algorithms. Automatic artifact removal using linear SVMs is a computationally cheap, robust, and extensible method [113]. In addition to ECG artifacts, ocular and muscular artifacts can be eliminated with ANN [114]. Recently the deep learning methods used for removal of artifact from the EEG signal [115,116].

9.3.5 Combined approach for artifact removal

To exploit the advantage of each method, recently, several researchers have opted to use a hybrid strategy, which is a combination of two or more methods. This categorization is based on how many individual algorithms are involved in the pre-processing pipeline. Popular single methods for EEG artifact removal include linear regression, adaptive filtering, WT, BSS, and empirical mode decomposition (EMD). On the contrary, the hybrid methods involve a combination of these single methods. Some major hybrid methods are discussed below.

9.3.5.1 Blind source separation and adaptive filtering

This hybrid technique is a combination of the BSS (BSS: ICA) and adaptive filtering. When it comes to eliminating artifacts, adaptive filtering is used to find them while BSS is responsible for locating them. Decomposing EEG signals into ICs is a task for ICA. The neuronal information contained in artificial ICs is preserved through further processing by an adaptive filter. In order to effectively remove artifacts from EEG data, combined adaptive filtering with ICA to create a hybrid method [43]. Cardiac artifact was removed using adaptive noise canceller and ICA [117]. ICA and adaptive noise cancellation (ANC), or ICA-ANC, are the foundation of this technique. As a means of determining the reference signal for ANC, ICA is applied to a small number of EEG signals. As the method only requires a small number of EEG channels and no synchronous ECG channel, it is well suited for use in portable BCIs [118]. An unsupervised and fully automatic method of detecting and eliminating visual artifacts using a combination of canonical correlation analysis and a multi-channel Wiener filter (ACCAMWF). Segments of artifacts can be automatically annotated using the spatial distribution entropy (SDE) and the spectral entropy (SE). After that, the clean EEG data are supplemented with neural signal extracted from artifact-contaminated data using the CCA algorithm [119].

9.3.5.2 Adaptive filtering and wavelet transform

To get rid of these ocular artifacts, [120] proposes a method that combines DWT and ANC, where the OA reference is derived from DWT decomposition and then used in the adaptive filter as reference. Ocular artefact removal using Multiresolution Analysis and Adaptive Filtering [MRAF] was used in [121]. To begin, discrete wavelet transform (DWT) is applied to the EEG signal in order to effectively help localize the epileptic region. These decomposed wavelet components are then subjected to multi-resolutional soft thresholding to smooth out any peaks and valleys. Low-frequency components representing physiological artifacts are also filtered out using adaptive filtering.

9.3.5.3 Technique of BSS and WT

Since the number of sources in ICA must be equal to the number of sources in the measurement, and the WT fails when the artifacts overlap in the spectral domain, a wavelet-ICA technique has been proposed in [81,122–124] to combine the good parts and avoid the bad ones. When processing EEG data, WICA is employed to filter out extraneous signals like EMG and to clean up single-channel recordings of the brain's electrical activity. After the EEG data has been recorded, the WT is used to decompose it, and the resulting resolution, which contains likely artifactual components, is then fed into an ICA algorithm of choice. Finally, using the preserved wavelet components and the disposed components, artifact-free reconstruction of the EEG signals is carried out.

An Automatic Wavelet Independent Component Analysis (AWICA) for an automatic artifact rejection from multichannel scalp EEG [125]. Firstly, the input EEG will be decomposed using the DWT to partition each channel of the original data set into four major bands of the brain activity. Each rhythm of each channel is represented by a Wavelet Component (WC). Each of the WCs will be channeled to ICA analysis to concentrate the artifacts into a few ICs. The resulting artifactual Wavelet Independent Components (WICs) are automatically detected and removed. Subsequently, the reconstruction of the clean signal involves two steps, which are inverse ICA and inverse DWT.

Wavelet is frequently used in conjunction with ICA for artifact removal in EEG [126]. Artifacts are identified using ICA to find the ICs. In order to obtain an EEG free of artifacts, we will flag the offending ICs as critical, denoise the signal using DWTs, and then rebuild the ICs using the corrected data. When trying to get usable neural signals out of artifact ICs, a lifting wavelet transform (LWT) can help [127]. Since ICA-LWT does not necessitate intricate computation, it can be used in real-time settings. An unsupervised eye blink artifact removal with modified multiscale sample entropy (mMSE), kurtosis, and wavelet-ICA is proposed by authors in [128,129]. After the decomposition of ICA, mMSE is computed for all the ICs. Normal distribution is used to detect the artifact. ICs with mMSE lower than the threshold will be marked for wavelet correction. The ICs computed will undergo the same process for Kurtosis. Finally, the marked ICs will be denoised using biorthogonal wavelets.

9.3.5.4 Technique of EMD and BSS

The EEG signals are first subjected to EMD in order to obtain IMFs; then, the IMFs are subjected to the BSS method, which detects and removes artifactual components [103]. By iteratively summing the remaining IMFs, the original signal can be reconstructed from scratch without any artifacts [130]. This is accomplished by multiplying the mixing matrix by the extracted IC components. Removing EMG and ocular artifact from EEG using EEMD in conjunction with ICA was first investigated in [131]. EEMD-ICA, EMD-CCA [132], EEMD-CCA [133–136], BSS-EMD, EEMD-SCICA [137], CEEMDAN-ICA-WTD [138], and multivariate empirical mode decomposition (MEMD) and CCA (MEMD-CCA) [139] used for removal of artifact from the EEG signal.

9.3.5.5 Adaptive filtering and EMD

The adaptive filtering and EMD are the two main components of this new method. Adaptive filter (using RLS algorithm) and empirical mode decomposition (EMD) are used to filter out ECG artifacts in EEG data [140].

9.3.5.6 Technique of BSS and SVM

BSS-SVM is a hybrid approach that can be used to further BSS's application. To begin, BSS algorithms are used to decompose the recorded EEG data into multiple components. The next step is to extract a variety of component features, including temporal, spatial, and statistical characteristics. The features are then fed into a linear SVM classifier to determine what kinds of artifact parts are present. Signals free of artifacts are reconstructed using the remaining components [22,25,141,142].

9.3.6 Summary of earlier methods used for EEG artifact removal

As number of artifact removal methods are discussed in previous section, for ease of observing all the above methods in terms of their methods, Performance parameters taken into consideration, way of acquiring EEG data are given in tabular form.

A summary of artifact removal techniques for filtering techniques is provided in Table 9.1. Decomposition techniques such as EMD and VMD and combined techniques that include EMD and VMD are provided in Table 9.2. Source separation algorithms and transformed methods such as WT, ST, and TWT are provided in Table 9.3. And recent machine learning techniques applied for artifact removal from EEG signals are provided in Table 9.4. It is observed that most of the authors collected EEG data from openly available databases for verification of their artifact removal algorithm. Some authors collect EEG data from the hospital, and some record EEG data using the recording machine available to them. The main problem with collecting EEG data is that appropriate data may not be available for removing particular types of artifacts. For recording appropriate EEG data to verify the efficacy of the artifact removal algorithm, a proper set of EEG recording machines is required, which is not cost-effective. In this regard, researchers depend on other sources for collecting EEG data. For the

Table 9.1 Filtering techniques utilized in EEG artifact removal process

Author	Method	Type of artifact	Performance measure	Data Base
C. S. Kim et al. [118]	Independent component analysis (ICA) and adaptive filter (AF)	Ocular	RMSE and CC	BCI Competition IV and Dataset IIIa of BCI Competition III
Bo Hua [143]	LMS (least mean square) based algorithm	Eye movement	SNR, MI, Coherence	Recorded
S. Blum et al. [144]	Artifact Subspace Reconstruction (ASR)	Eye blink	SNR, Sensitivity, Computation time	Data recorded using smart phone
EGLE BUTKEVIČİ UT̄ E et al. [145]	Baseline estimation and denoising with sparsity (BEADS) filter algorithm	Movement artifact	Pearson CC	Data recorded using smart phone
S. Kohli and A. J. Casson [146]	Moving averages and adaptive filtering	transcranial Direct Current Stimulation (tDCS) and transcranial alternating current artifact	SNR	Recorded
A. J. Mohammad Ali B. [117]	ICA and ANC	ECG artifact	RRMSE and frequency correlation	Mitsar amplifier and WinEEG software
M. Chavez et al. [147]	Surrogate-based artifact removal (SuBAR)	ocular and muscular artifacts	SNR and RRMSE	Recorded
M. Miao et al. [119]	CCA-MWF (ACCAMWF)	EOG	RMSE and MI	Semi-simulated EEG/EOG dataset(Mendelay)
S. Sharma and U. Satija [148]	DFT and ACMD	Ocular	CCN, SNR, AEM, AEMAX, MAXN, RDN, RDP, RMSE, SNRW, CCW, η	Mendeley database, MIT-BIH Polysomnographic database and EEGMAT database

(Continued)

Table 9.1 (Continued) Filtering techniques utilized in EEG artifact removal process

Author	Method	Type of artifact	Performance measure	Data Base
G. S. Spencer et al. [39]	RLAS	Motion artifact	RMSE	Recorded
A. Kilicarslan et al. [149]	Adaptive de-noising framework with H∞ adaptation rule	Motion artifact	PSD	Recorded
R. Ranjan et al. [150]	Modified-EMD and LoG filter	Motion artifact	ΔSNR, γ, mean absolute error in PSD of δ-band (MAEδ PSD), MI, percentage improvement in correlation [corr (%)], percentage improvement in coherence [coh(%)], power spectral distortion [PSDdis (%)] and execution time	Semi-simulated EEG data, mobile brain-body imaging (MoBI) real-time EEG dataset with BCI task and synthetic dataset

Table 9.2 Summery of decomposition techniques utilized for EEG artifact removal process

Author	Method	Type of artifact	Performance measure	Data Base
Shivam Sharma and Udit Satija [148]	DFT and ACMD	Ocular	CC, SNR, AE, AE MAX, MAXN, RDN, RDP, RMSE, SNRW, CCW, percentage reduction in coefficient of correlation (η), M I	Mendeley database, MIT-BIH polysomnographic database, EEGMAT database
Md S Hossain, et al. [151]	VMD, VMD-PCA, VMD-CCA	Motion artifact	SNR and percentage reduction in motion artifact	PhysioNet
R. Ranjan, et al. [150]	Modified EMD and optimized LoG filter	Motion artifact	SNR, signal to artifact gain coefficient, PSNR, SAR, MAE, MI, improvement in correlation, improvement in coherence, power spectral destruction and execution time	Semi-simulated EEG data contaminated with motion artifacts, mobile brain-body imaging (MoBI) real-time EEG data
Miao Shi, et al. [110]	GOSSA-VMD	Ocular	SNR, RMSE and CC	Semi-simulated EEG dataset
C. Kaur, et al. [152]	VMD-DWT and VMD-WPT		SNR, PSNR, and MSE.	Data collected from Hospital University Sains Malaysia (HUSM)
L. Chang, et al. [153]	MVMD-CCA		Accuracy and ITR	Simulated data using Matlab
M. Saini, et al. [109]	VMD in two stages denoted as VMD-I and VMD-II	Ocular	Sensitivity, NCC, SNR, MAE,MAX,NMAX,NRD, PRD, and RMSE	Mendeley database, MIT-BIH polysomnographic database and EEGMAT database
R. Gavas, et al. [112]	MVMD	Eye blink	SER, correlation, variance-based metric (V), percentage change in band power, classification accuracy	Synthetically generated EEG data, Covert Shift Dataset, Cog Beacon Dataset

(Continued)

Table 9.2 (Continued) Summery of decomposition techniques utilized for EEG artifact removal process

Author	Method	Type of artifact	Performance measure	Data Base
C. Dora and P. K. Biswal [154]	Modified VMD	ECG artifact	SAR, CF	MIT/BIH polysomnography data
Q. Li, et al. [138]	CEEMDAN-ICA-WTD	Ocular	RMSE	Recorded
C. Dora and P. K. Biswal [155]	VMD-based algorithm	Ocular	PSD	Capslpdb and sleep-edf of Physionet
A. Yadav and M. S. Choudhry [137]	EEMD and SCICA	Ocular	MI, CC and Coherence	physionet.org
Yan Liu, et al. [156]	NALSMEMD	Motion artifact	Percentage change in Artifact	physionet.org
X. Chen, et al. [133]	EEMD-CCA	Muscle artifact	SNR, RMS, RRMSE	Recorded and simulated data
A. Egambaram, et al. [132]	EMD and CCA	Eye blink	Accuracy (VMI), Error (VMI), CC, RMSE, Time of execution	Recorded
MD E. Alam, et al. [157]	EMD	Eye blink and power line noise	SNR	BCI2000
K. K. Dutta, et al. [158]	EEMD	Muscle artifact		Recorded
K. T Sweeney, et al. [135]	EEMD-CCA	Motion artifact	ΔSNR and percentage reduction in artifact λ	Recorded
S. Tavildar and A. Ashraf [159]	MEMD-CCA	Motion artifact	Correlation the percent reduction in artifacts λ	Recorded

Table 9.3 Methods of wavelet transform and blind source separation

Author	Method	Type of artifact	Performance measures	Source of EEG data
R. Mahajan and B. I. Morshed [128]	ICA and DWT	Eye blink	Sensitivity, specificity, agreement rate, MI and Cross Power spectral density	Recorded
Chi Zhang et al. [160]	DWT and ICA	EOG and EMG	Correlation scores and accuracy	Recorded
Charvi A. Majmudar et al. [161].	DWT	Ocular	TFA, Magnitude Square Coherence (MSC), CC, MI	Recorded
S. A. Gaikwad K.P.Paradeshi [162]	ICA(SOBI) and DWT	Ocular	SDR, Variance and RMSE	Recorded
R. Upadhyay et al. [163]	ICA-DOST	Ocular	By visual Inspection	Recorded and simulated
Xiaobai CAI and Junjun CHEN [164]	WT and HT	Eye blink	Visual inspection	Collected from Salk computational neurobiology Laboratory of California University
B. Somers and A. Bertrand [165]	CCA	Eye blink	SER, ARR, and SNR	Recorded
Mst. Jannatul Ferdous et al. [166]	Lifting WT	Eye blink	MSE and SAR	BCI Database
Sim Kuan Goh et al. [167]	ICA	Multiple	SNR, CSR	Synthesized EEG artifacts
Anwesha Khasnobish et al. [168]	KNN -DWT	Head movement artifact	SD	Recorded

(Continued)

Table 9.3 (Continued) Methods of wavelet transform and blind source separation

Author	Method	Type of artifact	Performance measures	Source of EEG data
Matteo Dora, David Holcman [169]	VT	Different kinds of artifact	NMSE, ΔR, ΔSNR	Semi-simulated EEG
Chi-Yuan Chang et al. [170]	ICA and ASR	Muscle, eye-blink, and lateral eye-movement activities	RMS	Driving simulator EEG data
S. R. Sreeja et al. [171]	MCAand KSVD	Eye blink	RMSE, SAR, CC, MI and MSE	Recorded
Nitesh S. Malan Shiru Sharma [172]	DTCWT	Ocular	RRMSE	Recordeded using NI LABVIEW 2015 Biosignal toolkit
Chong Yeh Sai et al. [113]	SVM and WICA	Eye blink	CC	Recorded and simulated
A. J.M. Ali Badamchizadeh et al. [117]	ICA-ANC	ECG artifact	RRMSE and frequency correlation	Recorded
Pranjali Gajbhiye et al. [173]	MTV, MWTV and DWT	Motion artifact	Difference in SNR and η	physionet
Md. Kafiul Islam Amir Rastegarnia [174]	SWT	Chewing, swallowing, eye blinks, subject movements, talking,	SNR, ΔSNR, ΔRMSE, ΔPSD, Δcorr, ΔSNDR	BCI competition-IV Scalp EEG Database: dataset-1, dataset-2a and dataset-2b
Soojin Lee et al. [175]	JBSS and quadrature regression and q-IVA	High-amplitude stimulation artifact	Root mean square (RMS), relative root-mean-squared error (RRMSE), correlation coefficient (CC), power deviation (Pdev)	Recorded and simulated

Author	Technique	Artifact	Metrics	Dataset
Mohamed F. Issa and Zoltan Juhasz [176]	WEICA	EOG artifact	Signal-to-noise ratio ΔSNR, RMSE, MSC, Percentage of artifact removal	Klados EEG dataset and Recorded dataset
Abhijit Bhattacharyya et al. [177]	TQWT	Cortical stimulation (CS)	MSE and CCI	Simulated data and Data Collected from Nancy University Hospital (CHU Nancy), France
Vandana Roy Shailja Shukla [178]	ICA,CCA,DWT and SWT	Motion artifact	ΔSNR, RMSE	Online open source interface
Young-Eun Lee et al. [179]	cICA with cIOL	Movement artifact	AUC(Area under the ROC curve) and SNR	Recorded
Chi-Yuan Chang et al. [180]	ASR	Eye blink and Muscle artifact	Retained power	Recorded
Dhanalekshmi P. Yedurkar, Shilpa P. Metkar [121]	DWT and MRAF	Ocular	Sensitivity, specificity, Accuracy and precision	CHB MIT scalp EEG dataset chb01 15
K. Jindal et al. [181]	FPIC and GLCT	Eye blink, EOG, muscular and other high frequency artifact	SNR, RMSE, NMAX and NRD	Simulated and recorded
Nikesh Bajaja et al. [182]	WPD	Muscle, motion and ocular artifact	MI, CC, PSD	Recorded
K. P. Paradeshi and U. D. Kolekar [183]	Enhanced WICA	Ocular	ADP, RMSE, PSNR, CC	Recorded
Sayedu Khasim Noorbasha, Gnanou Florence Sudha [184]	Ov-ASSA -ANC	Ocular	RRMSE, MAE	CHB-MIT

(Continued)

Table 9.3 (Continued) Methods of wavelet transform and blind source separation

Author	Method	Type of artifact	Performance measures	Source of EEG data
Ian McNulty et al. [185]	WT using the SURE Shrink algorithm with the hard thresholding	Ocular	NMSE and SNR	Bonn database
S. T. AUNG AND Y. W. WAT [186]	M-mDistEn	Motion artifact	Accuracy and p-values	PhysioNet Database
Pranjali Gajbhiye et al. [187]	WOSG filtering	Motion artifact	NMSE, ΔR, ΔSNR	Two publicly available databases
Zainab Jamil et al. [188]	ICA-DWT	Eye movement artifact	MI, Sensitivity and specificity	Recorded
S. Phadikar et al. [189]	WT With Heuristically Optimized threshold	Eye blink	CC, NMSE, SSIM	Recorded
M. Shahbakhti et al. [190]	SWT	Electrical shift and linear trend artifacts (ESLT)	CC, PSNR, NRMSE	Mendelay
Sridhar Chintala et al. [191]	Mixed step size normalized least mean fourth adaptive algorithm	Ocular	MSD	Recorded
Salim Çınar [192]	ICA-ANC	Ocular	RE, CC, SAR, SNR, Sensitivity, Specificity, and AUC	ERP-based Brain-Computer Interface (BCI) Records dataset
Christos Stergiadis et al. [193]	BSS	Ocular	Entropy, MIR, CC, Execution time	Real recorded data and semi simulated data

Reference	Method	Artifact	Metrics	Dataset
Ruisen Huang et al. [194]	DSMF	Motion artifact	SDR, NMSE	Recorded
Mary Judith A. et al. [195]	MD-SVD -ICA		SNR, PSNR, MSE	MIT EEG database
Velu Prabhakar Kumaravel et al. [196]	LOF and ASR	Newborn non stereotypical artifacts.	FTR, SME, F1 Score	Data collected from Neonatal Neuroimaging Unit CIMeC, University of Trento) and simulated data
Yuheng FENG et al. [197]	SSA-CCA	Muscle artifact	PSD and mean time cost	Semi simulated data
Sagar S. Motdhare, Dr. Garima Mathur [198]	SSA-MEMD	EMG artifact	Visual inspection	Recorded
A. K. Maddirala and K. C. Veluvolu [199]	SSA-CWT and k-means clustering	Eye blink	RRMSE, SNR, RMS, CC, artifact reduction ratio, MAE, precision and accuracy	Fatigue EEG database
J. Yedukondalu and L. D. Sharma [200]	CiSSA-DWT	EOG artifact	SAR, MAE, RRMSE, and CC	Dataset 2a from the BCI Competition IV

Table 9.4 Summery of machine learning technique in the field of artifact removal from EEG signal

Author	Method	Type of artifact	Performance measure	Data Base
C. Burger et al. [201]	ICA and WNN	Ocular	PSD, RMSE and frequency correlation	Simulated and recorded data collected from motor imagery test
W. Suna et al. [202]	ID-ResCNN model	EMG, ECG, EOG	SNR, RMSE	CHB-MIT Scalp EEG Database
S. K. Sahoo and S. K. Mohapatra [203]	EMCD and ODCN	Ocular	MAE, SNR	BCI competition IV database
R. Ghosh et al. [204]	kNN classifier and a LSTM network	Eyeblink and muscular	CC, SAR, MAE, NMSE, SSIM	Recorded
B. Yang et al. [116]	DLN	Ocular	PSD, RMSE and EEG classification accuracy	"Data sets I" for BCI Competition IV
M.H. Quazi et al. [205]	FLM optimization-based learning algorithm for neural network-enhanced adaptive filtering model	EMG, ECG, EOG	MSE, SNR	PhysioNet
X. Li et al. [206]	Discriminative model for joint OAC and feature learning	Ocular		Recorded
S. Behera and M. N. Mohanty [115]	WVFLN	Ocular	MSE, RMSE,	Mendeley database
S. Behera and M. N. Mohanty	RVFLN model	Ocular and cardiac artifact	MSE, NMSE, RE, GSAR, SNR, IQ, and INPS	Mendeley database

removal of cardiac artifacts, a proper data set is not available, so researchers create a synthesized version of an ECG artifact-contaminated EEG signal to verify the artifact removal algorithm. In [36], the authors recorded the cardiac artifact-contaminated EEG data from a small animal (a rat), and it is also practically very difficult to collect real-time EEG data. The researchers also rely on semi-simulated data. In the case of a semi-simulated dataset, the artifact is synthetically added to the clean EEG signal to form contaminated EEG data.

Another problem in the artifact removal process is the validation measure. It is not possible to verify the performance of an algorithm visually by observing the output signal obtained. Due to this, the researcher considers different types of measurement parameters to verify the performance of the artifact removal algorithm. It is observed from the tables that the performance measurement parameters are calculated by various authors. There is no fixed number of performance parameters that should be taken as a standard parameter in the case of artifact removal. It is also observed that in recent papers the authors considering a greater number of performance parameters for verifying their artifact removal algorithm.

9.4 PROPOSED TECHNIQUE

As the artifacts are unwanted and need to be removed prior to the information extraction, an automatic way of removing artifact is proposed in this paper. In the literature review section, some of the automatic artifact removal techniques are discussed. Most of the automatic EEG artifact removal methods are based on transforming the EEG signal to another domain. However, when the signal is transformed from the time domain to another domain it suffers from the problem of time-frequency resolution. To avoid the problem of time-frequency resolution the Fast Discrete S Transform (FDST) is proposed in this work.

9.4.1 Fast discrete S transform (FDST)

The general S transform given in equation (9.1) can be written as:

$$S_{eeg}\left(jT, \frac{n}{NT}\right) = \sum_{k=0}^{N-1} S_{eeg}(kT) \cdot w(kT, \zeta) \cdot e^{\frac{j2\pi kn}{N}} \qquad (9.1)$$

Where $S_{eeg}(kT)$ is the EEG signal, $k = [0, 1, 2, \dots N - 1]$ and N is the sample size of the EEG signal, T is the sampling interval of the signal considered. For implementing FDST the Fourier transform of EEG signal is obtained in the first step. Required kernel functions and the window function are calculated in the second step [207,208]. The bandpass filter is applied to the Fourier transform of the signal, and then inverse Fourier transform is applied to obtain S'_{eeg}. In the last step, FST is calculated for each kernel function, which are given in equation (9.2) and (9.3).

$$S_{eeg}\left[jT, \frac{3n}{4NT}\right] = \sum_{k=0}^{N-1} S'_{eeg}(kT) \cdot w\left(kT - T, \left|\frac{3n}{4NT}\right|\right) \cdot \phi^+\left(kT, \frac{3n}{4NT}\right) \qquad (9.2)$$

$$S_{eeg}\left[jT, -\frac{3n}{4NT}\right] = \sum_{k=0}^{N-1} S'_{eeg}(kT) \cdot w\left(kT - T, \left|\frac{3n}{4NT}\right|\right) \cdot \phi^-\left(kT, \frac{3n}{4NT}\right) \qquad (9.3)$$

From the above equations ϕ^+ and ϕ^- are the kernel functions, ζ is equivalent to v in DOST. $S'_{eeg}(kT)$ is obtained by taking inverse Fourier transform of the bandpass version of the original signal. The inverse of FDST is obtained by the equation (9.4)

$$S_{eeg}\left(\frac{n+1}{NT}\right) = \frac{1}{N}\left[\sum_{j=0}^{M-1} S_{eeg}\left(jT, \frac{n}{NT}\right) \cdot e^{-\frac{i2\pi jl}{M}}\right] \cdot \frac{1}{W\left(\frac{1}{NT}, \zeta\right)} \qquad (9.4)$$

Where $S_{eeg}\left(\frac{n+1}{NT}\right)$ is the Fourier transform of the EEG signal, and to get back the original signal we need to apply inverse Fourier transform. FDST and DOST use DFT for their operation and can be represented as:

$$DOST = \left(\sum_{i=1}^{k} B_i\right) DFT \qquad (9.5)$$

Where direct sum matrix B_i forms a block diagonal matrix. Each block in direct sum matrix is composed of a phase correction factor coming from $e^{i\pi\tau}$ and β dimensional inverse Fourier transform. The DOST basis functions are compact in frequency, but not in space. DOST coefficients carry symmetry property due to higher frequencies that are required in frequency, and in space DOST suffers from ringing effect. The problems that arise in the case of DOST and FDST can be overcome by DCST. The mathematical overview of DCST is given in the following subsection.

9.5 RESULT AND DISCUSSION

The proposed algorithm is tested by considering the clean EEG signal and two types of contaminated EEGs. The clean EEG is collected from the Mendeley data base [209], as shown in Figure 9.1.The length of the clean EEG signal is 6001, and its amplitude is within the range of $-30\,\mu V$ to $+30\,\mu V$. The clean EEG signal is considered here as a reference to validate

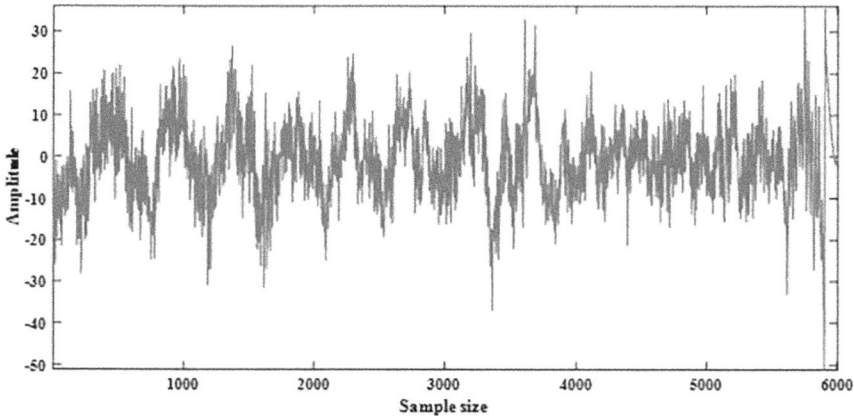

Figure 9.1 Clean EEG signal taken from Mendeley database.

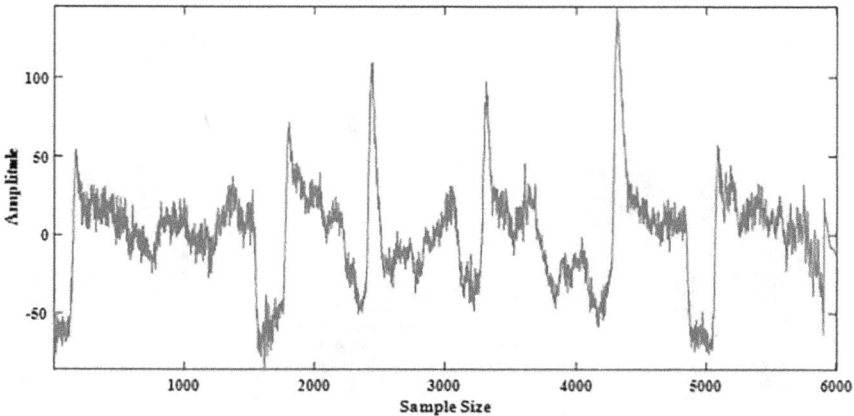

Figure 9.2 Contaminated EEG signal taken from Mendeley database.

the performance of the transform-based algorithm. One type of contaminated EEG collected from the Mendeley data base is the semi simulated EEG. In the case of semi simulated contaminated EEG, the artifactual EEG is obtained by adding the vertical and horizontal eye movement artifact by following the appropriate procedure and is only meant to verify the artifact removal algorithms. The contaminated signal is shown in Figure 9.2 and is of 6001 sample size. The amplitude of the contaminated EEG is within –60 μV to + 110 μV. By visual inspection it is clearly distinguished from the clean EEG signal. Another artifactual EEG signal collected from the PhysioNet database [210] is shown in Figure 9.3, and its amplitude is within the range of –600 μV to + 1600 μV. It is observed that at the time of occurrence of artifact, the signal amplitude drastically increased from

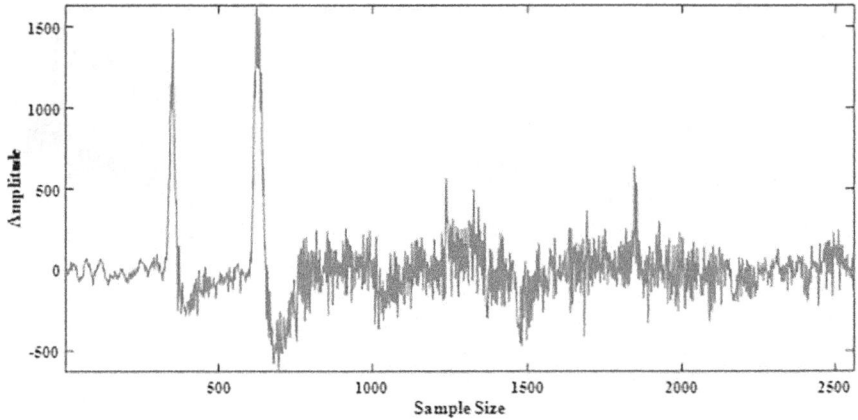

Figure 9.3 Artifactual EEG signal taken from PhysioNet database.

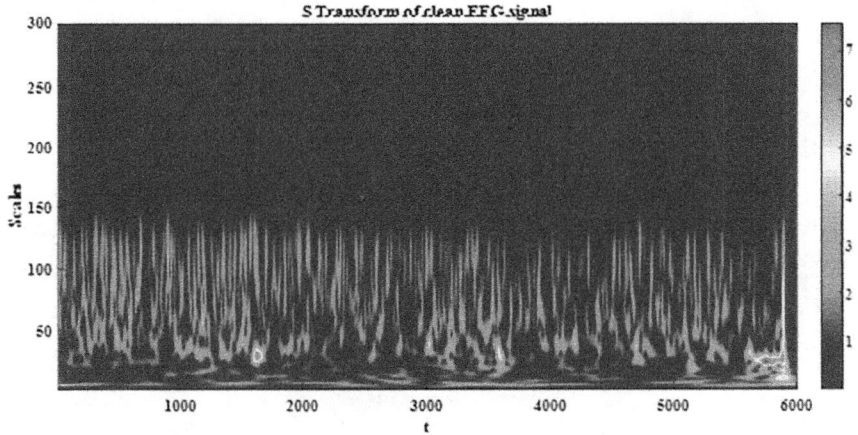

Figure 9.4 Fast DST of clean EEG signal.

its normal amplitude. Due to artifact the EEG signal shape is drastically changed, and it is removed by the proposed algorithm.

The FDST of clean EEG and contaminated EEG are shown in Figure 9.4 and Figure 9.5, respectively. From the Figure 9.4 and Figure 9.5 it is observed that at the time of occurrence of artifact, the phase change is more prominent. More phase changes are indicated by the red and yellow color.

Figure 9.6 shows the clean EEG signal, contaminated signal, and the signal after removal of artifact. It is observed that the recovered EEG signal amplitude is lesser than the amplitude of the contaminated EEG signal. Figure 9.7 shows the artifactual EEG signal collected from PhysioNet database and the recovered artifact-free EEG signal.

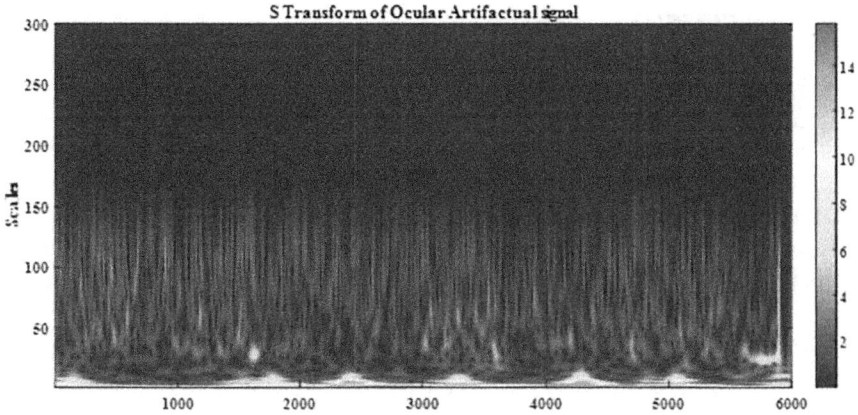

Figure 9.5 Fast DST of contaminated EEG signal.

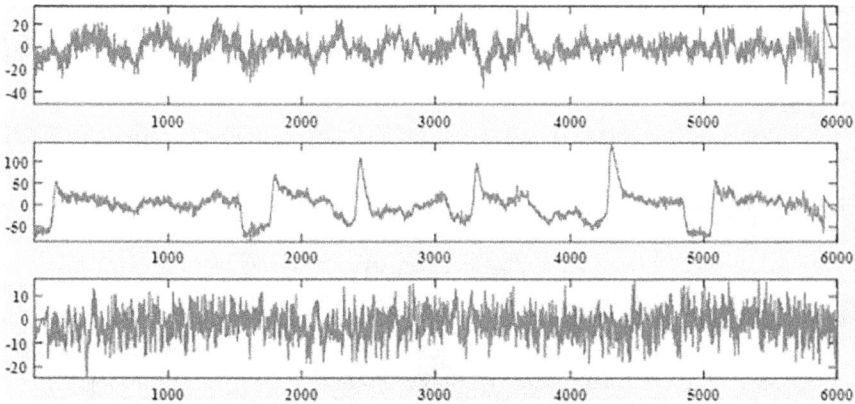

Figure 9.6 The clean EEG signal, contaminated EEG and the signal after removal of artifact.

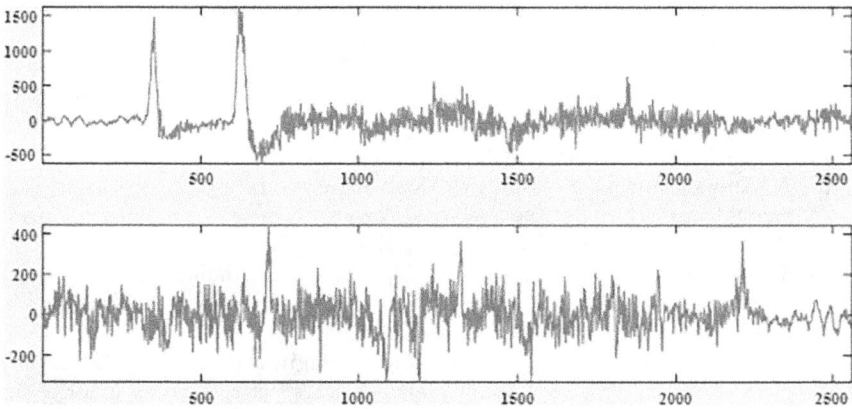

Figure 9.7 Artifactual EEG and signal after removal of artifact.

9.6 CONCLUSION

In this paper, we have covered some of the most common artifacts in EEG data as well as some of the methods currently in use to eliminate them. To this day, there is no universal method that can transform all contaminated EEG data into usable EEG data. To achieve desirable outcomes, it is common practice to employ multiple methods in sequence. It is common practice to use ICA-based algorithms as the next step after filtering in order to obtain clean EEG data. After that, various artifact removal strategies can be implemented, each tailored to the specific artifacts present in the dataset. Both WT and regression can be used to clean up EEG data that has been contaminated by electrocardiogram (ECG) or electrooculogram (EOG) artefacts, respectively. EOG and ECG artifact removal using machine learning techniques has shown remarkable improvement over the past three years. Muscle and body movement, as well as other extrinsic artifacts, should be minimized or eliminated whenever possible during the EEG recording process. Some artifacts, such as EOG and ECG artifacts, are unavoidable but can be eliminated with these techniques.

REFERENCES

[1] A. Subasi, "EEG signal classification using wavelet feature extraction and a mixture of expert model," *Expert Systems with Applications*, vol. 32, no. 4, pp. 1084–1093, 2007.

[2] T. Zhang, W. Chen, and M. Li, "Generalized Stockwell transform and SVD-based epileptic seizure detection in EEG using random forest," *Biocybernetics and Biomedical Engineering*, vol. 38, no. 3, pp. 519–534, 2018.

[3] W. Mumtaz, S. S. A. Ali, M. A. M. Yasin, and A. S. Malik, "A machine learning framework involving EEG-based functional connectivity to diagnose major depressive disorder (MDD)," *Medical & Biological Engineering & Computing*, vol. 56, pp. 233–246, 2018.

[4] U. R. Acharya, S. L. Oh, Y. Hagiwara, J. H. Tan, H. Adeli, and D. P. Subha, "Automated EEG-based screening of depression using deep convolutional neural network," *Computer Methods and Programs in Biomedicine*, vol. 161, pp. 103–113, 2018.

[5] A. Anuragi and D. S. Sisodia, "Alcohol use disorder detection using EEG Signal features and flexible analytical wavelet transform," *Biomedical Signal Processing and Control*, vol. 52, pp. 384–393, 2019.

[6] W. Mumtaz, N. Kamel, S. S. A. Ali, and A. S. Malik, "An EEG-based functional connectivity measure for automatic detection of alcohol use disorder," *Artificial Intelligence in Medicine*, vol. 84, pp. 79–89, 2018.

[7] R. Yuvaraj, U. Rajendra Acharya, and Y. Hagiwara, "A novel Parkinson's Disease Diagnosis Index using higher-order spectra features in EEG signals," *Neural Computing and Applications*, vol. 30, pp. 1225–1235, 2018.

[8] H. Yu, X. Lei, Z. Song, C. Liu, and J. Wang, "Supervised network-based fuzzy learning of EEG signals for Alzheimer's disease identification," *IEEE Transactions on Fuzzy Systems*, vol. 28, no. 1, pp. 60–71, 2019.

[9] P. Ofner and G. R. Müller-Putz, "Movement target decoding from EEG and the corresponding discriminative sources: A preliminary study," in *2015 37th Annual International Conference of the IEEE Engineering in Medicine and Biology Society (EMBC)*, 2015: IEEE, pp. 1468–1471.

[10] G. R. Müller-Putz *et al.*, "Towards non-invasive EEG-based arm/hand-control in users with spinal cord injury," in *2017 5th International Winter Conference on Brain-Computer Interface (BCI)*, 2017: IEEE, pp. 63–65.

[11] Y. Zeng *et al.*, "EEG-based identity authentication framework using face rapid serial visual presentation with optimized channels," *Sensors*, vol. 19, no. 1, p. 6, 2018.

[12] J. Chen, Z. Mao, W. Yao, and Y. Huang, "EEG-based biometric identification with convolutional neural network," *Multimedia Tools and Applications*, vol. 79, pp. 10655–10675, 2020.

[13] M. M. N. Mannan, M. Y. Jeong, and M. A. Kamran, "Hybrid ICA—Regression: Automatic identification and removal of ocular artifacts from electroencephalographic signals," *Frontiers in Human Neuroscience*, vol. 10, p. 193, 2016.

[14] M. M. N. Mannan, S. Kim, M. Y. Jeong, and M. A. Kamran, "Hybrid EEG—Eye tracker: Automatic identification and removal of eye movement and blink artifacts from electroencephalographic signal," *Sensors*, vol. 16, no. 2, p. 241, 2016.

[15] A. Delorme and S. Makeig, "EEGLAB: an open source toolbox for analysis of single-trial EEG dynamics including independent component analysis," *Journal of Neuroscience M*, vol. 134, no. 1, pp. 9–21, 2004.

[16] J. Iriarte *et al.*, "Independent component analysis as a tool to eliminate artifacts in EEG: a quantitative study," *Journal of Clinical Neurophysiology*, vol. 20, no. 4, pp. 249–257, 2003.

[17] S. Kotte and J. K. Dabbakuti, "Methods for removal of artifacts from EEG signal: A review," in *Journal of Physics: Conference Series*, vol. 1706, no. 1: IOP Publishing, p. 012093, 2020.

[18] A. Bisht, C. Kaur, and P. Singh, "Recent advances in artifact removal techniques for EEG signal processing," *Intelligent Communication, Control and Devices: Proceedings of ICICCD 2018*, pp. 385–392, 2020.

[19] W. Mumtaz, S. Rasheed, and A. Irfan, "Review of challenges associated with the EEG artifact removal methods," *Biomedical Signal Processing and Control*, vol. 68, p. 102741, 2021.

[20] A. Khosla, P. Khandnor, and T. Chand, "A comparative analysis of signal processing and classification methods for different applications based on EEG signals," *Biocybernetics and Biomedical Engineering*, vol. 40, no. 2, pp. 649–690, 2020.

[21] D. Mahmood, H. Nisar, and Y. V. Voon, "Removal of Physiological Artifacts from Electroencephalogram Signals: A Review and Case Study," in *2021 IEEE 9th Conference on Systems, Process and Control (ICSPC 2021)*, 2021: IEEE, pp. 141–146.

[22] X. Jiang, G.-B. Bian, and Z. Tian, "Removal of artifacts from EEG signals: a review," *Sensors*, vol. 19, no. 5, p. 987, 2019.

[23] M. K. Islam, A. Rastegarnia, and Z. Yang, "Les méthodes de détection et de rejet d'artefact de l'EEG de scalp: revue de littérature," *Neurophysiologie Clinique*, vol. 46, no. 4-5, pp. 287–305, 2016.

[24] M. M. N. Mannan, M. A. Kamran, and M. Y. Jeong, "Identification and removal of physiological artifacts from electroencephalogram signals: A review," *Ieee Access*, vol. 6, pp. 30630–30652, 2018.

[25] J. A. Urigüen and B. Garcia-Zapirain, "EEG artifact removal—state-of-the-art and guidelines," *Journal of Neural Engineering*, vol. 12, no. 3, p. 031001, 2015.

[26] M. K. Islam, A. Rastegarnia, and Z. Yang, "Methods for artifact detection and removal from scalp EEG: A review," *Neurophysiologie Clinique/Clinical Neurophysiology*, vol. 46, no. 4-5, pp. 287–305, 2016.

[27] A. Schlögl, C. Keinrath, D. Zimmermann, R. Scherer, R. Leeb, and G. Pfurtscheller, "A fully automated correction method of EOG artifacts in EEG recordings," *Clinical Neurophysiology*, vol. 118, no. 1, pp. 98–104, 2007.

[28] G. L. Wallstrom, R. E. Kass, A. Miller, J. F. Cohn, and N. A. Fox, "Automatic correction of ocular artifacts in the EEG: a comparison of regression-based and component-based methods," *International Journal of Psychophysiology*, vol. 53, no. 2, pp. 105–119, 2004.

[29] A. Q. Hamal and A. W. bin Abdul Rehman, "Artifact processing of epileptic EEG signals: an overview of different types of artifacts," in *2013 International Conference on Advanced Computer Science Applications and Technologies*, 2013: IEEE, pp. 358–361.

[30] R. J. Croft and R. J. Barry, "Removal of ocular artifact from the EEG: a review," *Neurophysiologie Clinique/Clinical Neurophysiology*, vol. 30, no. 1, pp. 5–19, 2000.

[31] I. I. Goncharova, D. J. McFarland, T. M. Vaughan, and J. R. Wolpaw, "EMG contamination of EEG: spectral and topographical characteristics," *Clinical Neurophysiology*, vol. 114, no. 9, pp. 1580–1593, 2003.

[32] S. D. Muthukumaraswamy, "High-frequency brain activity and muscle artifacts in MEG/EEG: a review and recommendations," *Frontiers in Human Neuroscience*, vol. 7, p. 138, 2013.

[33] X. Chen *et al.*, "Removal of muscle artifacts from the EEG: A review and recommendations," *IEEE Sensors Journal*, vol. 19, no. 14, pp. 5353–5368, 2019.

[34] R. D. O'Donnell, J. Berkhout, and W. R. Adey, "Contamination of scalp EEG spectrum during contraction of cranio-facial muscles," *Electroencephalography and Clinical Neurophysiology*, vol. 37, no. 2, pp. 145–151, 1974.

[35] K. Wang, W. Li, L. Dong, L. Zou, and C. Wang, "Clustering-constrained ICA for ballistocardiogram artifacts removal in simultaneous EEG-fMRI," *Frontiers in Neuroscience*, vol. 12, p. 59, 2018.

[36] C. Dai, J. Wang, J. Xie, W. Li, Y. Gong, and Y. Li, "Removal of ECG artifacts from EEG using an effective recursive least square notch filter," *IEEE Access*, vol. 7, pp. 158872–158880, 2019.

[37] P. J. Allen, G. Polizzi, K. Krakow, D. R. Fish, and L. Lemieux, "Identification of EEG events in the MR scanner: the problem of pulse artifact and a method for its subtraction," *Neuroimage*, vol. 8, no. 3, pp. 229–239, 1998.

[38] P. J. Allen, O. Josephs, and R. Turner, "A method for removing imaging artifact from continuous EEG recorded during functional MRI," *Neuroimage*, vol. 12, no. 2, pp. 230–239, 2000.

[39] G. S. Spencer, J. A. Smith, M. E. Chowdhury, R. Bowtell, and K. J. Mullinger, "Exploring the origins of EEG motion artefacts during simultaneous fMRI acquisition: Implications for motion artefact correction," *NeuroImage*, vol. 173, pp. 188–198, 2018.

[40] A. Islam, E. J. Esha, S. F. B. Ahmed, and M. K. Islam, "Study and Analysis of Motion Artifacts for Ambulatory EEG," *Article in International Journal of Electrical and Computer Engineering*, vol. 9, no. 4, 2021.

[41] J. T. Gwin, K. Gramann, S. Makeig, and D. P. Ferris, "Removal of movement artifact from high-density EEG recorded during walking and running," *Journal of Neurophysiology*, vol. 103, no. 6, pp. 3526–3534, 2010.

[42] M. H. Kutner, C. J. Nachtsheim, J. Neter, and W. Wasserman, *Applied linear regression models*. McGraw-Hill/Irwin New York, 2004.

[43] M. A. Klados, C. Papadelis, C. Braun, and P. D. Bamidis, "REG-ICA: a hybrid methodology combining blind source separation and regression techniques for the rejection of ocular artifacts," *Biomedical Signal Processing and Control*, vol. 6, no. 3, pp. 291–300, 2011.

[44] S. A. Hillyard and R. Galambos, "Eye movement artifact in the CNV," *Electroencephalography and Clinical Neurophysiology*, vol. 28, no. 2, pp. 173–182, 1970.

[45] J. L. Whitton, F. Lue, and H. Moldofsky, "A spectral method for removing eye movement artifacts from the EEG," *Electroencephalography and Clinical Neurophysiology*, vol. 44, no. 6, pp. 735–741, 1978.

[46] G. L. Wallstrom, R. E. Kass, A. Miller, and N. A. Fox, "in the EEG Using Bayesian Adaptive Regression Splines," *Case Studies in Bayesian Statistics*, vol. 6, p. 351, 2018.

[47] S. Romero, M. Mañanas, and M. J. Barbanoj, "Ocular reduction in EEG signals based on adaptive filtering, regression and blind source separation," *Annals of biomedical engineering*, vol. 37, pp. 176–191, 2009.

[48] T. Gasser, P. Ziegler, and W. F. Gattaz, "The deleterious effect of ocular artefacts on the quantitative EEG, and a remedy," *European Archives of Psychiatry and Clinical Neuroscience*, vol. 241, pp. 352–356, 1992.

[49] F. Ghaderi, S. K. Kim, and E. A. Kirchner, "Effects of eye artifact removal methods on single trial P300 detection, a comparative study," *Journal of Neuroscience Methods*, vol. 221, pp. 41–47, 2014.

[50] K. T. Sweeney, T. E. Ward, and S. F. McLoone, "Artifact removal in physiological signals—Practices and possibilities," *IEEE Transactions on Information Technology in Biomedicine*, vol. 16, no. 3, pp. 488–500, 2012.

[51] P. Sadasivan and D. N. Dutt, "ANC schemes for the enhancement of EEG signals in the presence of EOG artifacts," *Computers and Biomedical Research*, vol. 29, no. 1, pp. 27–40, 1996.

[52] D. Hagemann and E. Naumann, "The effects of ocular artifacts on (lateralized) broadband power in the EEG," *Clinical Neurophysiology*, vol. 112, no. 2, pp. 215–231, 2001.

[53] S. Narasimhan and D. N. Dutt, "Application of LMS adaptive predictive filtering for muscle artifact (noise) cancellation from EEG signals," *Computers & Electrical Engineering*, vol. 22, no. 1, pp. 13–30, 1996.

[54] F. Bartoli and S. Cerutti, "An optimal linear filter for the reduction of noise superimposed to the EEG signal," *Journal of Biomedical Engineering*, vol. 5, no. 4, pp. 274–280, 1983.

[55] V. Ingle, S. Kogon, and D. Manolakis, *Statistical and adaptive signal processing*. Artech, 2005.

[56] S. Puthusserypady and T. Ratnarajah, "H/sup/spl infin//adaptive filters for eye blink artifact minimization from electroencephalogram," *IEEE Signal Processing Letters*, vol. 12, no. 12, pp. 816–819, 2005.

[57] A. G. Correa, E. Laciar, H. Patiño, and M. Valentinuzzi, "Artifact removal from EEG signals using adaptive filters in cascade," in *Journal of Physics: Conference Series*, vol. 90, no. 1: IOP Publishing, p. 012081, 2007.

[58] B. Jervis, M. Thomlinson, C. Mair, J. Lopez, and M. Garcia, "Residual ocular artefact subsequent to ocular artefact removal from the electroencephalogram," *IEE Proceedings-Science, Measurement and Technology*, vol. 146, no. 6, pp. 293–298, 1999.

[59] P. He, G. Wilson, and C. Russell, "Removal of ocular artifacts from electroencephalogram by adaptive filtering," *Medical and Biological Engineering and Computing*, vol. 42, pp. 407–412, 2004.

[60] C. Marque, C. Bisch, R. Dantas, S. Elayoubi, V. Brosse, and C. Perot, "Adaptive filtering for ECG rejection from surface EMG recordings," *Journal of Electromyography and Kinesiology*, vol. 15, no. 3, pp. 310–315, 2005.

[61] M. Milanesi *et al.*, "Multichannel techniques for motion artifacts removal from electrocardiographic signals," in *2006 International Conference of the IEEE Engineering in Medicine and Biology Society*, 2006: IEEE, pp. 3391–3394.

[62] J. Mateo, E. M. Sánchez-Morla, and J. Santos, "A new method for removal of powerline interference in ECG and EEG recordings," *Computers & Electrical Engineering*, vol. 45, pp. 235–248, 2015.

[63] R. Sameni, M. B. Shamsollahi, and C. Jutten, "Model-based Bayesian filtering of cardiac contaminants from biomedical recordings," *Physiological Measurement*, vol. 29, no. 5, p. 595, 2008.

[64] J. J. Kierkels, J. Riani, J. W. Bergmans, and G. J. Van Boxtel, "Using an eye tracker for accurate eye movement artifact correction," *IEEE Transactions on Biomedical Engineering*, vol. 54, no. 7, pp. 1256–1267, 2007.

[65] B. Somers, T. Francart, and A. Bertrand, "A generic EEG artifact removal algorithm based on the multi-channel Wiener filter," *Journal of Neural Engineering*, vol. 15, no. 3, p. 036007, 2018.

[66] P. Diniz, "Adaptive Filtering: Algorithms and Practical Implementation. Springer," *New York, NY, USA*, 2008.

[67] G. Welch, "An Introduction to the Kalman Filter" "Univ. of North Carolina," http://www.cs.unc.edu/~welch/media/pdf/kalman_intro.pdf, 2006.

[68] M. Izzetoglu, P. Chitrapu, S. Bunce, and B. Onaral, "Motion artifact cancellation in NIR spectroscopy using discrete Kalman filtering," *Biomedical Engineering Online*, vol. 9, pp. 1–10, 2010.

[69] S. Seyedtabaii and L. Seyedtabaii, "Kalman filter based adaptive reduction of motion artifact from photoplethysmographic signal," *World Acad. Sci. Eng. Technol*, vol. 37, pp. 173–176, 2008.

[70] F. Morbidi, A. Garulli, D. Prattichizzo, C. Rizzo, and S. Rossi, "Application of Kalman filter to remove TMS-induced artifacts from EEG recordings," *IEEE Transactions on Control Systems Technology*, vol. 16, no. 6, pp. 1360–1366, 2008.

[71] J. V. Candy, *Bayesian signal processing: classical, modern, and particle filtering methods*. John Wiley & Sons, 2016.

[72] C.-M. Ting, S.-H. Salleh, Z. M. Zainuddin, and A. Bahar, "Spectral estimation of nonstationary EEG using particle filtering with application to event-related desynchronization (ERD)," *IEEE Transactions on Biomedical Engineering*, vol. 58, no. 2, pp. 321–331, 2010.

[73] S. Nakagome, T. P. Luu, J. A. Brantley, and J. L. Contreras-Vidal, "Prediction of EMG envelopes of multiple terrains over-ground walking from EEG signals using an unscented Kalman filter," in *2017 IEEE International Conference on Systems, Man, and Cybernetics (SMC)*, 2017: IEEE, pp. 3175–3178.

[74] T. P. Luu, Y. He, S. Nakagame, J. Gorges, K. Nathan, and J. L. Contreras-Vidal, "Unscented Kalman filter for neural decoding of human treadmill walking from non-invasive electroencephalography," in *2016 38th Annual International Conference of the IEEE Engineering in Medicine and Biology Society (EMBC)*, 2016: IEEE, pp. 1548–1551.

[75] M. Yakoubi, R. Hamdi, and M. B. Salah, "EEG enhancement using extended Kalman filter to train multi-layer perceptron," *Biomedical Engineering: Applications, Basis and Communications*, vol. 31, no. 01, p. 1950005, 2019.

[76] C.-M. Ting, S.-H. Salleh, Z. M. Zainuddin, and A. Bahar, "Artifact removal from single-trial ERPs using non-Gaussian stochastic volatility models and particle filter," *IEEE Signal Processing Letters*, vol. 21, no. 8, pp. 923–927, 2014.

[77] V. Fox, J. Hightower, L. Liao, D. Schulz, and G. Borriello, "Bayesian filtering for location estimation," *IEEE Pervasive Computing*, vol. 2, no. 3, pp. 24–33, 2003.

[78] S. Choi, A. Cichocki, H.-M. Park, and S.-Y. Lee, "Blind source separation and independent component analysis: A review," *Neural Information Processing-Letters and Reviews*, vol. 6, no. 1, pp. 1–57, 2005.

[79] A. Delorme, T. Sejnowski, and S. Makeig, "Enhanced detection of artifacts in EEG data using higher-order statistics and independent component analysis," *Neuroimage*, vol. 34, no. 4, pp. 1443–1449, 2007.

[80] A. Mognon, J. Jovicich, L. Bruzzone, and M. Buiatti, "ADJUST: An automatic EEG artifact detector based on the joint use of spatial and temporal features," *Psychophysiology*, vol. 48, no. 2, pp. 229–240, 2011.

[81] I. Winkler, S. Brandl, F. Horn, E. Waldburger, C. Allefeld, and M. Tangermann, "Robust artifactual independent component classification for BCI practitioners," *Journal of Neural Engineering*, vol. 11, no. 3, p. 035013, 2014.

[82] T.-P. Jung, S. Makeig, M. Westerfield, J. Townsend, E. Courchesne, and T. J. Sejnowski, "Removal of eye activity artifacts from visual event-related potentials in normal and clinical subjects," *Clinical Neurophysiology*, vol. 111, no. 10, pp. 1745–1758, 2000.

[83] Y. Li, Z. Ma, W. Lu, and Y. Li, "Automatic removal of the eye blink artifact from EEG using an ICA-based template matching approach," *Physiological Measurement*, vol. 27, no. 4, p. 425, 2006.

[84] N. Mammone and F. C. Morabito, "Enhanced automatic artifact detection based on independent component analysis and Renyi's entropy," *Neural Networks*, vol. 21, no. 7, pp. 1029–1040, 2008.

[85] S. P. Fitzgibbon, D. M. Powers, K. J. Pope, and C. R. Clark, "Removal of EEG noise and artifact using blind source separation," *Journal of Clinical Neurophysiology*, vol. 24, no. 3, pp. 232–243, 2007.

[86] C. J. James and O. J. Gibson, "Temporally constrained ICA: an application to artifact rejection in electromagnetic brain signal analysis," *IEEE Transactions on Biomedical Engineering*, vol. 50, no. 9, pp. 1108–1116, 2003.

[87] A. Delorme, J. Palmer, J. Onton, R. Oostenveld, and S. Makeig, "Independent EEG sources are dipolar," *PloS One*, vol. 7, no. 2, p. e30135, 2012.

[88] K. Ting, P. Fung, C. Chang, and F. Chan, "Automatic correction of artifact from single-trial event-related potentials by blind source separation using second order statistics only," *Medical Engineering & Physics*, vol. 28, no. 8, pp. 780–794, 2006.

[89] W. De Clercq, A. Vergult, B. Vanrumste, W. Van Paesschen, and S. Van Huffel, "Canonical correlation analysis applied to remove muscle artifacts from the electroencephalogram," *IEEE Transactions on Biomedical Engineering*, vol. 53, no. 12, pp. 2583–2587, 2006.

[90] D. M. Vos et al., "Removal of muscle artifacts from EEG recordings of spoken language production," *Neuroinformatics*, vol. 8, pp. 135–150, 2010.

[91] X. Yong, R. K. Ward, and G. E. Birch, "Artifact removal in EEG using morphological component analysis," in *2009 IEEE International Conference on Acoustics, Speech and Signal Processing*, 2009: IEEE, pp. 345–348.

[92] P. Berg and M. Scherg, "Dipole modelling of eye activity and its application to the removal of eye artefacts from the EEG and MEG," *Clinical Physics and Physiological Measurement*, vol. 12, no. A, p. 49, 1991.

[93] S. Casarotto, A. M. Bianchi, S. Cerutti, and G. A. Chiarenza, "Principal component analysis for reduction of ocular artefacts in event-related potentials of normal and dyslexic children," *Clinical Neurophysiology*, vol. 115, no. 3, pp. 609–619, 2004.

[94] P. S. Kumar, R. Arumuganathan, K. Sivakumar, and C. Vimal, "Removal of ocular artifacts in the EEG through wavelet transform without using an EOG reference channel," *Int. J. Open Problems Compt. Math*, vol. 1, no. 3, pp. 188–200, 2008.

[95] D. Safieddine et al., "Removal of muscle artifact from EEG data: comparison between stochastic (ICA and CCA) and deterministic (EMD and wavelet-based) approaches," *EURASIP Journal on Advances in Signal Processing*, vol. 2012, no. 1, pp. 1–15, 2012.

[96] V. Krishnaveni, S. Jayaraman, L. Anitha, and K. Ramadoss, "Removal of ocular artifacts from EEG using adaptive thresholding of wavelet coefficients," *Journal of Neural Engineering*, vol. 3, no. 4, p. 338, 2006.

[97] M. Unser and A. Aldroubi, "A review of wavelets in biomedical applications," *Proceedings of the IEEE*, vol. 84, no. 4, pp. 626–638, 1996.

[98] S. Slobounov, M. Hallett, C. Cao, and K. Newell, "Modulation of cortical activity as a result of voluntary postural sway direction: an EEG study," *Neuroscience Letters*, vol. 442, no. 3, pp. 309–313, 2008.

[99] G. Inuso, F. La Foresta, N. Mammone, and F. C. Morabito, "Wavelet-ICA methodology for efficient artifact removal from Electroencephalographic recordings," in *2007 international joint conference on neural networks*, 2007: IEEE, pp. 1524–1529.

[100] P. S. Kumar, R. Arumuganathan, K. Sivakumar, and C. Vimal, "An adaptive method to remove ocular artifacts from EEG signals using wavelet transform," *J. Appl. Sci. Res*, vol. 5, no. 7, pp. 711–745, 2009.

[101] M. K. I. Molla, M. R. Islam, T. Tanaka, and T. M. Rutkowski, "Artifact suppression from EEG signals using data adaptive time domain filtering," *Neurocomputing*, vol. 97, pp. 297–308, 2012.

[102] Z. Wu and N. E. Huang, "Ensemble empirical mode decomposition: a noise-assisted data analysis method," *Advances in Adaptive Data Analysis*, vol. 1, no. 01, pp. 1–41, 2009.

[103] B. Mijović, M. De Vos, I. Gligorijević, J. Taelman, and S. Van Huffel, "Source separation from single-channel recordings by combining empirical-mode decomposition and independent component analysis," *IEEE Transactions on Biomedical Engineering*, vol. 57, no. 9, pp. 2188–2196, 2010.

[104] M. G. Keshava and K. Z. Ahmed, "Correction of ocular artifacts in EEG signal using empirical mode decomposition and cross-correlation," *Research Journal of Biotechnology*, vol. 9, no. 12, pp. 21–26, 2014.

[105] M. Guarascio and S. Puthusserypady, "Automatic minimization of ocular artifacts from electroencephalogram: A novel approach by combining Complete EEMD with Adaptive Noise and Renyi's Entropy," *Biomedical Signal Processing and Control*, vol. 36, pp. 63–75, 2017.

[106] B. Yang, T. Zhang, Y. Zhang, W. Liu, J. Wang, and K. Duan, "Removal of electrooculogram artifacts from electroencephalogram using canonical correlation analysis with ensemble empirical mode decomposition," *Cognitive Computation*, vol. 9, pp. 626–633, 2017.

[107] K. Dragomiretskiy and D. Zosso, "Variational mode decomposition," *IEEE Transactions on Signal Processing*, vol. 62, no. 3, pp. 531–544, 2013.

[108] T. Dutta, U. Satija, B. Ramkumar, and M. S. Manikandan, "A novel method for automatic modulation classification under non-Gaussian noise based on variational mode decomposition," in *2016 twenty second national conference on communication (NCC)*, 2016: IEEE, pp. 1–6.

[109] M. Saini and U. Satija, "An effective and robust framework for ocular artifact removal from single-channel EEG signal based on variational mode decomposition," *IEEE Sensors Journal*, vol. 20, no. 1, pp. 369–376, 2019.

[110] M. Shi, C. Wang, W. Zhao, X. Zhang, Y. Ye, and N. Xie, "Removal of ocular artifacts from electroencephalo-graph by improving variational mode decomposition," *China Communications*, vol. 19, no. 2, pp. 47–61, 2022.

[111] N. ur Rehman and H. Aftab, "Multivariate variational mode decomposition," *IEEE Transactions on Signal Processing*, vol. 67, no. 23, pp. 6039–6052, 2019.

[112] R. Gavas, D. Jaiswal, D. Chatterjee, V. Viraraghavan, and R. K. Ramakrishnan, "Multivariate variational mode decomposition based approach for blink removal from EEG signal," in *2020 IEEE International Conference on Pervasive Computing and Communications Workshops (PerCom Workshops)*, 2020: IEEE, pp. 1–6.

[113] C. Y. Sai, N. Mokhtar, H. Arof, P. Cumming, and M. Iwahashi, "Automated classification and removal of EEG artifacts with SVM and wavelet-ICA," *IEEE Journal of Biomedical and Health Informatics*, vol. 22, no. 3, pp. 664–670, 2017.

[114] S. Behera and M. N. Mohanty, "Classification of EEG signal using SVM," in *Advances in Electrical Control and Signal Systems: Select Proceedings of AECSS 2019*, 2020: Springer, pp. 859–869.

[115] S. Behera and M. N. Mohanty, "Artifact removal using deep WVFLN for brain signal diagnosis through IoMT," *Measurement: Sensors*, vol. 24, p. 100465, 2022.

[116] B. Yang, K. Duan, C. Fan, C. Hu, and J. Wang, "Automatic ocular artifacts removal in EEG using deep learning," *Biomedical Signal Processing and Control*, vol. 43, pp. 148–158, 2018.

[117] A. Jafarifarmand and M. A. Badamchizadeh, "Real-time cardiac artifact removal from EEG using a hybrid approach," in *2018 International Conference BIOMDLORE*, 2018: IEEE, pp. 1–5.

[118] C. S. Kim, J. Sun, D. Liu, Q. Wang, and S. G. Paek, "Removal of ocular artifacts using ICA and adaptive filter for motor imagery-based BCI," *IEEE/CAA journal of automatica sinica*, 2017.

[119] M. Miao, W. Hu, B. Xu, J. Zhang, J. J. Rodrigues, and V. H. C. de Albuquerque, "Automated CCA-MWF algorithm for unsupervised identification and removal of EOG artifacts from EEG," *IEEE Journal of Biomedical and Health Informatics*, vol. 26, no. 8, pp. 3607–3617, 2021.

[120] H. Peng *et al.*, "Removal of ocular artifacts in EEG—An improved approach combining DWT and ANC for portable applications," *IEEE Journal of Biomedical and health informatics*, vol. 17, no. 3, pp. 600–607, 2013.

[121] D. P. Yedurkar and S. P. Metkar, "Multiresolution approach for artifacts removal and localization of seizure onset zone in epileptic EEG signal," *Biomedical Signal Processing and Control*, vol. 57, p. 101794, 2020.

[122] J. Lin and A. Zhang, "Fault feature separation using wavelet-ICA filter," *NDT & e International*, vol. 38, no. 6, pp. 421–427, 2005.

[123] B. Azzerboni, M. Carpentieri, F. La Foresta, and F. Morabito, "Neural-ICA and wavelet transform for artifacts removal in surface EMG," in *2004 IEEE International Joint Conference on Neural Networks (IEEE Cat. No. 04CH37541)*, 2004, vol. 4: IEEE, pp. 3223–3228.

[124] S. Calcagno, F. La Foresta, and M. Versaci, "Independent component analysis and discrete wavelet transform for artifact removal in biomedical signal processing," *American Journal of Applied Sciences*, vol. 11, no. 1, p. 57, 2014.

[125] N. Mammone, F. La Foresta, and F. C. Morabito, "Automatic artifact rejection from multichannel scalp EEG by wavelet ICA," *IEEE Sensors Journal*, vol. 12, no. 3, pp. 533–542, 2011.

[126] N. K. Al-Qazzaz, S. Hamid Bin Mohd Ali, S. A. Ahmad, M. S. Islam, and J. Escudero, "Automatic artifact removal in EEG of normal and demented individuals using ICA–WT during working memory tasks," *Sensors*, vol. 17, no. 6, p. 1326, 2017.

[127] S. Jirayucharoensak and P. Israsena, "Automatic removal of EEG artifacts using ICA and lifting wavelet transform," in *2013 International Computer Science and Engineering Conference (ICSEC)*, 2013: IEEE, pp. 136–139.

[128] R. Mahajan and B. I. Morshed, "Unsupervised eye blink artifact denoising of EEG data with modified multiscale sample entropy, kurtosis, and wavelet-ICA," *IEEE Journal of Biomedical and Health Informatics*, vol. 19, no. 1, pp. 158–165, 2014.

[129] S. Behera and M. N. Mohanty, "A statistical approach for ocular artifact removal in brain signals," in *2018 2nd International Conference on Data Science and Business Analytics (ICDSBA)*, 2018: IEEE, pp. 500–503.

[130] X. Chen, Z. J. Wang, and M. McKeown, "Joint blind source separation for neurophysiological data analysis: Multiset and multimodal methods," *IEEE Signal Processing Magazine*, vol. 33, no. 3, pp. 86–107, 2016.

[131] P. He, M. Kahle, G. Wilson, and C. Russell, "Removal of ocular artifacts from EEG: a comparison of adaptive filtering method and regression method using simulated data," in *2005 IEEE Engineering in Medicine and Biology 27th Annual Conference*, 2006: IEEE, pp. 1110–1113.

[132] A. Egambaram, N. Badruddin, V. S. Asirvadam, E. Fauvet, C. Stolz, and T. Begum, "Automated and online eye blink artifact removal from electro-encephalogram," in *2019 IEEE International Conference on Signal and Image Processing Applications (ICSIPA)*, 2019: IEEE, pp. 159–163.

[133] X. Chen, Q. Chen, Y. Zhang, and Z. J. Wang, "A novel EEMD-CCA approach to removing muscle artifacts for pervasive EEG," *IEEE Sensors Journal*, vol. 19, no. 19, pp. 8420–8431, 2018.

[134] X. Chen, C. He, and H. Peng, "Removal of muscle artifacts from single-channel EEG based on ensemble empirical mode decomposition and multiset canonical correlation analysis," *Journal of Applied Mathematics*, vol. 2014, 2014.

[135] K. T. Sweeney, S. F. McLoone, and T. E. Ward, "The use of ensemble empirical mode decomposition with canonical correlation analysis as a novel artifact removal technique," *IEEE Transactions on Biomedical Engineering*, vol. 60, no. 1, pp. 97–105, 2012.

[136] G. Wang, C. Teng, K. Li, Z. Zhang, and X. Yan, "The removal of EOG artifacts from EEG signals using independent component analysis and multivariate empirical mode decomposition," *IEEE Journal of Oiomedical and Health Informatics*, vol. 20, no. 5, pp. 1301–1308, 2015.

[137] A. Yadav and M. S. Choudhry, "A new approach for ocular artifact removal from EEG signal using EEMD and SCICA," *Cogent Engineering*, vol. 7, no. 1, p. 1835146, 2020.

[138] Q. Li, L. Wei, Y. Xu, and B. Yang, "Ocular Artifact Removal Algorithm of Single Channel EEG Based on CEEMDAN-ICA-WTD," in *2021 IEEE 6th International Conference on Signal and Image Processing (ICSIP)*, 2021: IEEE, pp. 451–455.

[139] X. Chen, X. Xu, A. Liu, M. J. McKeown, and Z. J. Wang, "The use of multivariate EMD and CCA for denoising muscle artifacts from few-channel EEG recordings," *IEEE Transactions on Instrumentation and Measurement*, vol. 67, no. 2, pp. 359–370, 2017.

[140] X. Navarro, F. Porée, and G. Carrault, "ECG removal in preterm EEG combining empirical mode decomposition and adaptive filtering," in *2012 IEEE international conference on acoustics, speech and signal processing (ICASSP)*, 2012: IEEE, pp. 661–664.

[141] L. Shoker, S. Sanei, and J. Chambers, "Artifact removal from electro-encephalograms using a hybrid BSS-SVM algorithm," *IEEE Signal Processing Letters*, vol. 12, no. 10, pp. 721–724, 2005.

[142] S. Halder *et al.*, "Online artifact removal for brain-computer interfaces using support vector machines and blind source separation," *Computational Intelligence and Neuroscience*, vol. 2007, 2007.

[143] B. Hua, "Improved adaptive filtering based artifact removal from EEG signals," in *2020 13th International Congress on Image and Signal Processing, BioMedical Engineering and Informatics (CISP-BMEI)*, 2020: IEEE, pp. 424–428.

[144] S. Blum, N. S. Jacobsen, M. G. Bleichner, and S. Debener, "A Riemannian modification of artifact subspace reconstruction for EEG artifact handling," *Frontiers in Human Neuroscience*, vol. 13, p. 141, 2019.

[145] E. Butkevičiūtė *et al.*, "Removal of movement artefact for mobile EEG analysis in sports exercises," *IEEE Access*, vol. 7, pp. 7206–7217, 2019.

[146] S. Kohli and A. J. Casson, "Removal of gross artifacts of transcranial alternating current stimulation in simultaneous EEG monitoring," *Sensors*, vol. 19, no. 1, p. 190, 2019.

[147] M. Chavez, F. Grosselin, A. Bussalb, F. D. V. Fallani, and X. Navarro-Sune, "Surrogate-based artifact removal from single-channel EEG," *IEEE Transactions on Neural Systems and Rehabilitation Engineering*, vol. 26, no. 3, pp. 540–550, 2018.

[148] S. Sharma and U. Satija, "Automated Ocular Artifacts Removal Framework Based on Adaptive Chirp Mode Decomposition," *IEEE Sensors Journal*, vol. 22, no. 6, pp. 5806–5814, 2022.

[149] A. Kilicarslan and J. L. C. Vidal, "Characterization and real-time removal of motion artifacts from EEG signals," *Journal of Neural Engineering*, vol. 16, no. 5, p. 056027, 2019.

[150] R. Ranjan, B. C. Sahana, and A. K. Bhandari, "Motion artifacts suppression from EEG signals using an adaptive signal denoising method," *IEEE Transactions on Instrumentation and Measurement*, vol. 71, pp. 1–10, 2022.

[151] M. S. Hossain *et al.*, "Motion artifacts correction from EEG and fNIRS signals using novel multiresolution analysis," *IEEE Access*, vol. 10, pp. 29760–29777, 2022.

[152] C. Kaur, A. Bisht, P. Singh, and G. Joshi, "EEG Signal denoising using hybrid approach of Variational Mode Decomposition and wavelets for depression," *Biomedical Signal Processing and Control*, vol. 65, p. 102337, 2021.

[153] L. Chang, R. Wang, and Y. Zhang, "Decoding SSVEP patterns from EEG via multivariate variational mode decomposition-informed canonical correlation analysis," *Biomedical Signal Processing and Control*, vol. 71, p. 103209, 2022.

[154] C. Dora and P. K. Biswal, "Correlation-based ECG artifact correction from single channel EEG using modified variational mode decomposition," *Computer Methods and Programs in Biomedicine*, vol. 183, p. 105092, 2020.

[155] C. Dora and P. K. Biswal, "An improved algorithm for efficient ocular artifact suppression from frontal EEG electrodes using VMD," *Biocybernetics and Biomedical Engineering*, vol. 40, no. 1, pp. 148–161, 2020.

[156] Y. Liu, F. An, X. Lang, and Y. Dai, "Remove motion artifacts from scalp single channel EEG based on noise assisted least square multivariate empirical mode decomposition," in *2020 13th International Congress on Image and Signal Processing, BioMedical Engineering and Informatics (CISP-BMEI)*, 2020: IEEE, pp. 568–573.

[157] M. E. Alam and B. Samanta, "Performance evaluation of empirical mode decomposition for EEG artifact removal," in *ASME International Mechanical Engineering Congress and Exposition*, 2017, vol. 58387: American Society of Mechanical Engineers, p. V04BT05A024.

[158] K. K. Dutta, K. Venugopal, and S. A. Swamy, "Removal of muscle artifacts from EEG based on ensemble empirical mode decomposition and classification of seizure using machine learning techniques," in *2017 International Conference on Inventive Computing and Informatics (ICICI)*, 2017: IEEE, pp. 861–866.

[159] S. Tavildar and A. Ashrafi, "Application of multivariate empirical mode decomposition and canonical correlation analysis for EEG motion artifact removal," in *2016 Conference on Advances in Signal Processing (CASP)*, 2016: IEEE, pp. 150–154.

[160] C. Zhang *et al.*, "Automatic artifact removal from electroencephalogram data based on a priori artifact information," *BioMed research international*, vol. 2015, 2015.

[161] C. A. Majmudar, R. Mahajan, and B. I. Morshed, "Real-time hybrid ocular artifact detection and removal for single channel EEG," in *2015 IEEE International Conference on Electro/Information Technology (EIT)*, 2015: IEEE, pp. 330–334.

[162] S. Gaikwad and K. Paradeshi, "Design of effective algorithm for removal of ocular artifact from multichannel EEG Signal using ICA and wavelet method," *International Journal of Computer Science and Information Technologies*, vol. 7, no. 3, pp. 1531–1535, 2016.

[163] R. Upadhyay, P. Padhy, and P. K. Kankar, "EEG artifact removal and noise suppression by discrete orthonormal S-transform denoising," *Computers & Electrical Engineering*, vol. 53, pp. 125–142, 2016.

[164] C. Xiaobai and C. Junjun, "A Method for Blink Artifact Detection and Removal with Wavelet Transform and Hilbert Transform," in *International Conference on Biological Engineering and Pharmacy 2016 (BEP 2016)*, 2016: Atlantis Press, pp. 104–109.

[165] B. Somers and A. Bertrand, "Removal of eye blink artifacts in wireless EEG sensor networks using reduced-bandwidth canonical correlation analysis," *Journal of neural engineering*, vol. 13, no. 6, p. 066008, 2016.

[166] M. J. Ferdous, M. S. Ali, M. E. Hamid, and M. K. I. Molla, "Eyeblink Artifact Suppression from EEG Signal using Lifting Wavelet Transform."

[167] S. K. Goh, H. A. Abbass, K. C. Tan, A. Al-Mamun, C. Wang, and C. Guan, "Automatic EEG artifact removal techniques by detecting influential independent components," *IEEE Transactions on Emerging Topics in Computational Intelligence*, vol. 1, no. 4, pp. 270–279, 2017.

[168] A. Khasnobish, K. Chakravarty, D. Chatterjee, and A. Sinha, "Wavelet based head movement artifact removal from electrooculography signals," in *2017 IEEE International Conference on Acoustics, Speech and Signal Processing (ICASSP)*, 2017: IEEE, pp. 984–988.

[169] M. Dora and D. Holcman, "Adaptive single-channel EEG artifact removal with applications to clinical monitoring," *IEEE Transactions on Neural Systems and Rehabilitation Engineering*, vol. 30, pp. 286–295, 2022.

[170] C.-Y. Chang, S.-H. Hsu, L. Pion-Tonachini, and T.-P. Jung, "Evaluation of artifact subspace reconstruction for automatic EEG artifact removal," in *2018 40th Annual International Conference of the IEEE Engineering in Medicine and Biology Society (EMBC)*, 2018: IEEE, pp. 1242–1245.

[171] S. Sreeja, R. R. Sahay, D. Samanta, and P. Mitra, "Removal of eye blink artifacts from EEG signals using sparsity," *IEEE Journal of Biomedical and Health Informatics*, vol. 22, no. 5, pp. 1362–1372, 2017.

[172] N. S. Malan and S. Sharma, "Removal of Ocular Atrifacts from Single Channel EEG Signal Using DTCWT with Quantum Inspired Adaptive Threshold," in *2018 2nd International Conference on Biomedical Engineering (IBIOMED)*, 2018: IEEE, pp. 94–99.

[173] P. Gajbhiye, R. K. Tripathy, A. Bhattacharyya, and R. B. Pachori, "Novel approaches for the removal of motion artifact from EEG recordings," *IEEE Sensors Journal*, vol. 19, no. 22, pp. 10600–10608, 2019.

[174] M. K. Islam and A. Rastegarnia, "Probability mapping based artifact detection and wavelet denoising based artifact removal from scalp EEG for BCI applications," in *2019 IEEE 4th International Conference on Computer and Communication Systems (ICCCS)*, 2019: IEEE, pp. 243–247.

[175] S. Lee, M. J. McKeown, Z. J. Wang, and X. Chen, "Removal of high-voltage brain stimulation artifacts from simultaneous EEG recordings," *IEEE Transactions on Biomedical Engineering*, vol. 66, no. 1, pp. 50–60, 2018.

[176] M. F. Issa and Z. Juhasz, "Improved EOG artifact removal using wavelet enhanced independent component analysis," *Brain Sciences*, vol. 9, no. 12, p. 355, 2019.

[177] A. Bhattacharyya et al., "A multi-channel approach for cortical stimulation artefact suppression in depth EEG signals using time-frequency and spatial filtering," *IEEE Transactions on Biomedical Engineering*, vol. 66, no. 7, pp. 1915–1926, 2018.

[178] V. Roy and S. Shukla, "Designing efficient blind source separation methods for EEG motion artifact removal based on statistical evaluation," *Wireless Personal Communications*, vol. 108, pp. 1311–1327, 2019.

[179] Y.-E. Lee, N.-S. Kwak, and S.-W. Lee, "A real-time movement artifact removal method for ambulatory brain-computer interfaces," *IEEE Transactions on Neural Systems and Rehabilitation Engineering*, vol. 28, no. 12, pp. 2660–2670, 2020.

[180] C.-Y. Chang, S.-H. Hsu, L. Pion-Tonachini, and T.-P. Jung, "Evaluation of artifact subspace reconstruction for automatic artifact components removal in multi-channel EEG recordings," *IEEE Transactions on Biomedical Engineering*, vol. 67, no. 4, pp. 1114–1121, 2019.
[181] K. Jindal, R. Upadhyay, and H. S. Singh, "Application of hybrid GLCT-PICA de-noising method in automated EEG artifact removal," *Biomedical Signal Processing and Control*, vol. 60, p. 101977, 2020.
[182] N. Bajaj, J. R. Carrión, F. Bellotti, R. Berta, and A. De Gloria, "Automatic and tunable algorithm for EEG artifact removal using wavelet decomposition with applications in predictive modeling during auditory tasks," *Biomedical Signal Processing and Control*, vol. 55, p. 101624, 2020.
[183] K. Paradeshi and U. Kolekar, "Ocular artifact suppression in multichannel EEG using dynamic segmentation and enhanced wICA," *IETE Journal of Research*, vol. 68, no. 4, pp. 2683–2696, 2022.
[184] S. K. Noorbasha and G. F. Sudha, "Removal of EOG artifacts from single channel EEG–an efficient model combining overlap segmented ASSA and ANC," *Biomedical Signal Processing and Control*, vol. 60, p. 101987, 2020.
[185] I. McNulty *et al.*, "Analysis of Artifacts Removal Techniques in EEG Signals for Energy-Constrained Devices," in *2021 IEEE International Midwest Symposium on Circuits and Systems (MWSCAS)*, 2021: IEEE, pp. 515–519.
[186] S. T. Aung and Y. Wongsawat, "Analysis of EEG signals contaminated with motion artifacts using multiscale modified-distribution entropy," *IEEE Access*, vol. 9, pp. 33911–33921, 2021.
[187] P. Gajbhiye, N. Mingchinda, W. Chen, S. C. Mukhopadhyay, T. Wilaiprasitporn, and R. K. Tripathy, "Wavelet domain optimized Savitzky–Golay filter for the removal of motion artifacts from EEG recordings," *IEEE Transactions on Instrumentation and Measurement*, vol. 70, pp. 1–11, 2020.
[188] Z. Jamil, A. Jamil, and M. Majid, "Artifact removal from EEG signals recorded in non-restricted environment," *Biocybernetics and Biomedical Engineering*, vol. 41, no. 2, pp. 503–515, 2021.
[189] S. Phadikar, N. Sinha, and R. Ghosh, "Automatic eyeblink artifact removal from EEG signal using wavelet transform with heuristically optimized threshold," *IEEE Journal of Biomedical and Health Informatics*, vol. 25, no. 2, pp. 475–484, 2020.
[190] M. Shahbakhti *et al.*, "SWT-kurtosis based algorithm for elimination of electrical shift and linear trend from EEG signals," *Biomedical Signal Processing and Control*, vol. 65, p. 102373, 2021.
[191] S. Chintala, J. Thangaraj, and D. R. Edla, "Mixed step size normalized least mean fourth adaptive algorithm for artifact elimination from raw EEG signals," *Biomedical Signal Processing and Control*, vol. 65, p. 102392, 2021.
[192] S. Çınar, "Design of an automatic hybrid system for removal of eye-blink artifacts from EEG recordings," *Biomedical Signal Processing and Control*, vol. 67, p. 102543, 2021.

[193] C. Stergiadis, V.-D. Kostaridou, and M. A. Klados, "Which BSS method separates better the EEG Signals? A comparison of five different algorithms," *Biomedical Signal Processing and Control*, vol. 72, p. 103292, 2022.

[194] R. Huang, K. Qing, D. Yang, and K.-S. Hong, "Real-time motion artifact removal using a dual-stage median filter," *Biomedical Signal Processing and Control*, vol. 72, p. 103301, 2022.

[195] A. M. Judith, S. B. Priya, and R. K. Mahendran, "Artifact Removal from EEG signals using Regenerative Multi-Dimensional Singular Value Decomposition and Independent Component Analysis," *Biomedical Signal Processing and Control*, vol. 74, p. 103452, 2022.

[196] V. P. Kumaravel, E. Farella, E. Parise, and M. Buiatti, "NEAR: An artifact removal pipeline for human newborn EEG data," *Developmental Cognitive Neuroscience*, vol. 54, p. 101068, 2022.

[197] Y. Feng, Q. Liu, A. Liu, R. Qian, and X. Chen, "A Novel SSA-CCA Framework forMuscle Artifact Removal from Ambulatory EEG," *Virtual Reality & Intelligent Hardware*, vol. 4, no. 1, pp. 1–21, 2022.

[198] S. S. Motdhare and G. Mathur, "An Experimental Analysis on EMG Artifact Removal Methods from EEG Signal Records," *Mathematical Statistician and Engineering Applications*, vol. 71, no. 1, pp. 72–78-72–78, 2022.

[199] A. K. Maddirala and K. C. Veluvolu, "SSA with CWT and k-means for eye-blink artifact removal from single-channel EEG signals," *Sensors*, vol. 22, no. 3, p. 931, 2022.

[200] J. Yedukondalu and L. D. Sharma, "Circulant Singular Spectrum Analysis and Discrete Wavelet Transform for Automated Removal of EOG Artifacts from EEG Signals," *Sensors*, vol. 23, no. 3, p. 1235, 2023.

[201] C. Burger and D. J. Van Den Heever, "Removal of EOG artefacts by combining wavelet neural network and independent component analysis," *Biomedical Signal Processing and Control*, vol. 15, pp. 67–79, 2015.

[202] W. Sun, Y. Su, X. Wu, and X. Wu, "A novel end-to-end 1D-ResCNN model to remove artifact from EEG signals," *Neurocomputing*, vol. 404, pp. 108–121, 2020.

[203] S. K. Sahoo and S. K. Mohapatra, "Recognition of Ocular Artifacts in EEG Signal through a Hybrid Optimized Scheme," *BioMed Research International*, vol. 2022, 2022.

[204] R. Ghosh, S. Phadikar, N. Deb, N. Sinha, P. Das, and E. Ghaderpour, "Automatic Eye-blink and Muscular Artifact Detection and Removal from EEG Signals using k-Nearest Neighbour Classifier and Long Short-Term Memory Networks," *IEEE Sensors Journal*, 2023.

[205] M. Quazi and S. Kahalekar, "Artifacts removal from EEG signal: FLM optimization-based learning algorithm for neural network-enhanced adaptive filtering," *Biocybernetics and Biomedical Engineering*, vol. 37, no. 3, pp. 401–411, 2017.

[206] X. Li, C. Guan, H. Zhang, and K. K. Ang, "Discriminative ocular artifact correction for feature learning in EEG analysis," *IEEE Transactions on biomedical engineering*, vol. 64, no. 8, pp. 1906–1913, 2016.

[207] R. A. Brown and R. Frayne, "A fast discrete S-transform for biomedical signal processing," in *2008 30th Annual International Conference of the IEEE Engineering in Medicine and Biology Society*, 2008: IEEE, pp. 2586–2589.

[208] W. Yao, Q. Tang, Z. Teng, Y. Gao, and H. Wen, "Fast S-transform for time-varying voltage flicker analysis," *IEEE Transactions on Instrumentation and Measurement*, vol. 63, no. 1, pp. 72–79, 2013.

[209] M. A. Klados and P. D. Bamidis, "A semi-simulated EEG/EOG dataset for the comparison of EOG artifact rejection techniques," *Data in brief*, vol. 8, pp. 1004–1006, 2016.

[210] I. Silva and G. B. Moody, "An open-source toolbox for analysing and processing physionet databases in matlab and octave," *Journal of Open Research Software*, vol. 2, no. 1, 2014.

Chapter 10

Analysis of neural network and neuromorphic computing with hardware

A survey

Manish Bhardwaj[1], Kailash Nath Tripathi[2], Yogendra Narayan Prajapati[3], and Analp Pathak[4]

[1]Department of Computer Science and Information Technology, KIET Group of Institutions, Delhi-NCR, Ghaziabad, India
[2]Department of AIML, ISBM College of Engineering, Pune, India
[3]Department of Computer Science and Engineering, Ajay Kumar Engineering, Ghaziabad, India
[4]Department of Information Technology, KIET Group of Institutions, Delhi-NCR, Ghaziabad, India

10.1 INTRODUCTION

In this study, the author presents a thorough overview of the field of neuromorphic computing, including topics such as objectives, neuron/synapse concepts, procedures and education, deployments, advances in hardware, and supplementary materials and systems [1]. By providing a comprehensive and historical overview of the topic, we hope to stimulate future research and serve as a jumping-off point for interested newcomers [2–4].

For decades, the goal of computer scientists has been to construct a system that can perceive the world at a rate greater than that of a person, and the von Neumann architecture has emerged as the undisputed gold standard for this kind of system. While parallels to the human brain are inescapable, the vastly different organizational structure, power requirements, and processing capabilities of the two systems underline the limitations of both architectures [5]. This begs the obvious question of whether or not artificial neural networks (ANNs) can be designed to perform as well as the human brain.

Paralleling von Neumann systems, neuromorphic computing has developed in recent years. In 1990, the term "neuromorphic computing" was coined by Mead; "neuromorphic" refers to a type of very large scale integration (VLSI) that uses analog components to simulate organic neural networks [6]. These days, the phrase also refers to implementations that make use of or are inspired by neural networks that aren't necessarily based on biology.

DOI: 10.1201/9781003398066-10

These neuromorphic designs are unique in that they are highly linked and parallel, need little energy, and co-localize storing and function. The impending end of Moore's law, the rising power requirements of Dennard scaling, and the poor connectivity between processor and memory, known as the von Neumann bottleneck, have all drawn more attention to neuromorphic designs, which are interesting in their own right. When compared to conventional von Neumann designs, neuromorphic computers have the ability to do complicated calculations more quickly, with less power consumption, and in a smaller physical footprint [7]. Taking advantage of neuromorphic systems in hardware development is strongly recommended due to these features.

Furthermore, machine learning is a driving force behind the growing popularity of neuromorphic computing. This method has the potential to significantly enhance learning efficiency on specific tasks [8–10]. The focus shifts from the architecture benefits of research in artificial intelligence to its prospective operational benefits, with the hope that programs capable of online, real-time learning like that of biological brains can be developed, shown in Figure 10.1. Modern machine learning technique implementations may find neuromorphic structures to be the most perfect platform.

Researchers from many disciplines, including materials science, neuroscience, electrical engineering, computer engineering, and computer science, are all represented in the neuromorphic computing community. Neuromorphic technologies make use of substances with attributes comparable to those of biological neural systems; therefore, neuroscientists explore, produce, and characterize new materials for application in these devices [11]. Researchers in the field of neuroscience employ neuromorphic systems to imitate and analyze biological neural systems, and they disseminate information about new findings from their investigations that may have computational applications [12–14]. Device-level analog, digital, mixed analog/digital, and non-traditional circuits are all tools of the trade for electrical and computer engineers as they develop and

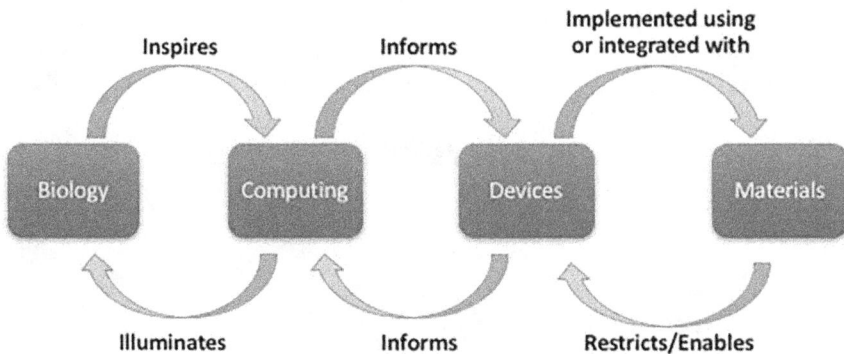

Figure 10.1 Various related fields with neuromorphic research.

implement novel devices, systems, and modes of communication. Inspired by biological systems and machine learning, computer scientists and engineers aim to create novel network models that can be trained and/or learn on their own. Aside from creating neuromorphic computing systems, they also create the enabling software for those systems to be used in practice.

Both von Neumann and Turing discussed brain-inspired systems in the 1950s, so the idea of employing specialized hardware to construct neurally inspired machines is as old as both computer science and computer engineering [15]. For quite some time, it has been a goal of the computing community to simulate the neuronal circuitry of the human brain. There have been several significant advances in ANNs, artificial intelligence (AI), and machine learning as a result of this quest. While ANNs and neuroscience are not the primary emphasis of this study, the creation of non-von Neumann hardware to model ANNs and biological brain systems is. Based on incentives mentioned in the literature, we analyze many reasons why neuromorphic systems have been developed over the years [16]. The exponential growth in the number of published studies related to neuromorphic computing over the past decade is shown in Figure 10.2. The evolution of 10 of the most influential reasons given in the literature for using a neuromorphic approach is depicted. Each of these top 10 inspirations has been mentioned in the literature at least 15 times, making them strong contenders for inclusion in the list.

Much of the pioneering accounts in neuromorphic computing was motivated by the creation of hardware that could conduct parallel processes, taking cues from the reported complexity in neurons and glia but running on a single chip. Although parallel architectures existed, neuromorphic systems differentiated themselves by emphasizing a large number of relatively simple processing components (often neurons) with a high density of connections between them (typically synapses) [17–20]. The intrinsic parallelism of neuromorphic systems was the most common justification for bespoke practical systems in early efforts on neuromorphic computing.

The ability to do computations quickly was another motivation for early neuromorphic and neural network hardware implementations. Particularly,

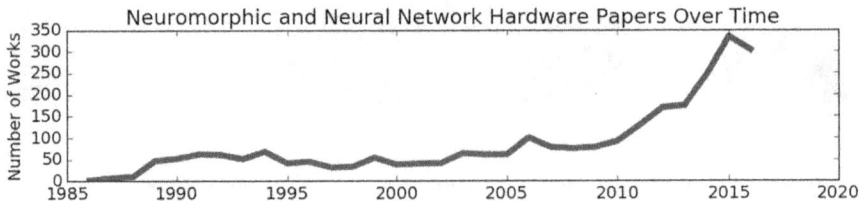

Figure 10.2 Graphical representation of research in neuromorphic and neural network hardware.

early system developers stressed that it was possible to achieve much faster neural network computation with custom chips than was possible with traditional von Neumann architectures, in part by extracting their natural computation, as described above, but also by adding custom hardware to finalize neural-style mathematical calculations [21–25]. This early emphasis on speed suggested that neuromorphic devices could be used as boosters for machine learning or neural network-style activities in the future.

One of the primary drivers of early neuromorphic systems was the need for real-time performance. In applications like real-time control, real-time digital image reconstruction, and autonomous robot control, the devices' natural computation and computational speed allowed neural network simulations to be completed more quickly than in implementations on von Neumann architectures [26–28]. The demand for quicker computing in these instances was driven more by the performance requirements of the underlying applications than by research into the topologies of neural networks. This is why we have separated it as a driving force behind the evolution of machine learning from the pursuit of speed and parallelism.

Due to various inherent single points of failure, both in the parallelized representation and in the potential adaptation or self-healing capabilities observed in ANN interpretations in software, developers have begun to consider neural networks as a natural template for hardware design [29]. In the past and now, these traits mattered for making new hardware implementations, as imperfections in both the created and used devices are possible due to component and technique variance.

The possibility for exceptionally low power performance is the most common reason given in the current research and discussions on neuromorphic devices in the cited papers [30]. The human brain is our primary source of inspiration, but it only uses approximately 20 watts of electricity and is capable of incredibly complicated computations and jobs. From its inception, the goal of developing neuromorphic devices with comparably low power consumption has been a driving force for neuromorphic computing. This goal has just recently emerged as a key motive [31–33].

During this century of neuromorphic research, the development of portable devices with the computational power of neural networks but with a minimal resource requirement (in terms of device size) has emerged as a driving force [34]. As the prevalence of integrated systems and microprocessors has increased, so has the need for designs that consume little space, have different contexts, and use minimal energy.

Recently, reduced prevalence has been the driving force for the creation of neuromorphic devices. As can be seen in Figure 10.2, this is the primary driving force for neuromorphic computing. Major drivers for the advancement of neuromorphic systems continue to be their inherent parallelism, real-time performance, speed in both operation and training, and tiny device footprint [35]. During this time, a few other motivations gained traction; one example is the proliferation of methods that employ neural network-style architectures (i.e., architectures made up of neuron and synapse-like components) due to their fault-tolerance characteristics or reliability in the face of hardware errors. In light of the usage of innovative materials for building neuromorphic systems, this has been an increasingly popular rationale in recent years.

The research of neuroscience has also been a driving force behind the development of neuromorphic systems in the last decade. Numerous neuroscience-driven efforts, such as the European Union's Human Brain Project [36], have relied on the development of custom domain adaptation because it is impractical to simulate genuine neural behavior on a conventional supercomputer due to its size, speed, and energy requirements. Therefore, in order to efficiently run significant neuroscience simulations, it is necessary to develop one's own neuromorphic implementations. Similarly, scalability has emerged as a major driver in the development of neuromorphic systems [37]. Most extensive neuromorphic projects discuss cascading their devices to achieve many neurons and synapses. However, most of the other specified justifications are related to the study of neuromorphic systems as a potential complementary architecture in the beyond Moore's law computing landscape, which is a common motivation that is not explicitly given in Figure 10.2. Many neuroscientists do not believe neuromorphic structures will substitute von Neumann architectures, but "building a better computer" is a driving force behind the creation of neuromorphic devices. This motivation is fairly all-encompassing, covering problems with conventional computers such as the impending end of Moore's law and Dennard scaling. The von Neumann bottleneck, which occurs in von Neumann architectures due to the separation of memory and processing and the performance gap between processing and memory technologies in present-day systems, is another driving force for the creation of neuromorphic computing. By colocating memory and computing, neuromorphic systems reduce the effects of the von Neumann bottleneck [38].

One of the driving forces behind the development of neuromorphic systems in recent years is the promise of online learning, or the capacity to quickly and effectively adjust to new circumstances within a task as they arise [39]. Though online learning mechanisms are currently poorly defined, many neuromorphic systems feature online learning components that could be used to accomplish learning tasks in an unsupervised, low-power fashion. Systems that can process and analyze this data in an unsupervised, online manner will be crucial in future computing platforms because of the exponential expansion of information collection in recent years. In addition, as our knowledge of biological brains expands, we should be able to design more effective mechanisms for online learning, with neuromorphic computing serving as an ideal platform for doing so.

10.2 MODELS OF RESEARCH

Which model of neural networks to select is a crucial concern in neuro-morphic computing. The components of the network, their functions, and their relationships with one another are all spelled out in the neural network model. Inspired by biological brain networks, basic types of neural networks include neurons and synapses. One must establish models for each component (e.g., neurons and synapses) while constructing the neural system model, with each component model dictating the behavior of its respective constituents [40].

How does one decide which model to use? It's possible that the selected model is inspired by a certain field of use. If, for the purposes of a neuroscience investigation, a more rapid simulation of biological brains is required than can be achieved using conventional von Neumann structures, then a model that is biologically credible and/or realistic is required for the neuromorphic device. A neuromorphic system that uses neural networks with convolution may be the best option if the application calls for very accurate picture recognition. Modifying the model itself is an option.

A specific tool or material has limitations and/or properties. For instance, models of spiking neural networks are best suited to the properties of systems based on memristors, such as spike-timing sensitive plasticity-like mechanismsn. In many other situations, it is unclear which model to use or how complex it should be.

Neuromorphic and neural network equipment technologies have been used to implement many different kinds of models [41–45]. The models can be primarily computationally generated or primarily biologically inspired. ANN models, rather than biological brains, serve as inspiration for the latter. In this section, we'll go over the various neuron models, synapse models, and network models that have been implemented in neuromorphic systems, and we'll highlight some of the most important articles for each.

10.2.1 Models of neurons

There are three main parts that make up a neuron in the body: the cell body, the axon, and the dendrites. Neurons typically (but not always) use their axon to send their signals away from the cell. Typically, dendrites are the site of intake to neurons, and they are responsible for relaying that knowledge to the cell body, though this is not always the case. Neurons can take in data from many other neurons via chemical or electrical signals. A presynaptic is the connection between the synapses and the axon of two neurons that provides for the communication of data or signals between them. Upon receiving a signal from another

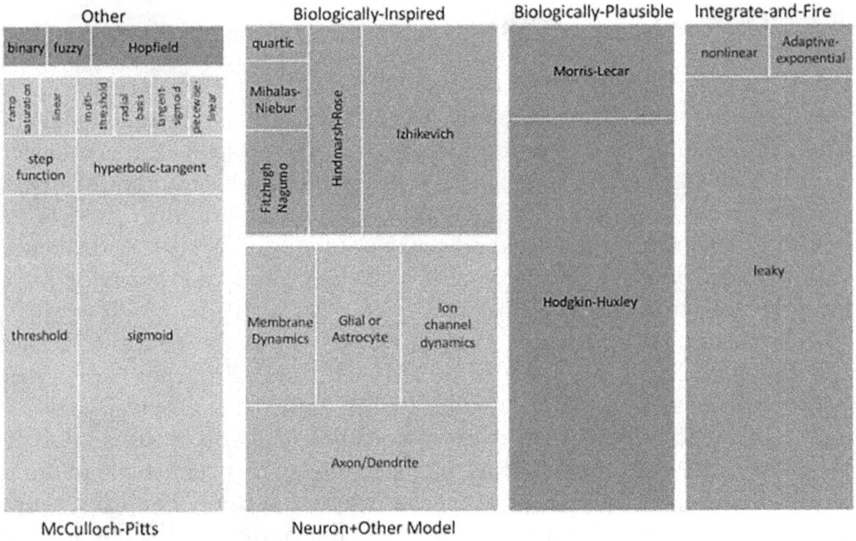

Figure 10.3 Hardware implementation in neuron models.

neuron via a synapse, a neuron will often begin to charge up as a result of a shift in the power supply throughout the neuron's cell membrane. When the voltage inside a neuron reaches a certain threshold, the neuron "fires," or produces a muscle contraction that travels along the axon and modulates the charge on neighboring neurons via synapses. Different neuron models in neuromorphic systems may use different techniques to accomplish the same ideas of charge transport and firing to influence neighboring neurons. Similarly, axons and synapses are not often implemented in artificial simulations that are not scientifically realistic (i.e., models inspired by neuroscience instead of copying neuroscience).

The various neuron model implementations are shown in Figure 10.3. There are five main types of neuron models provided:

To be biologically credible, a model must explicitly simulate the varieties of behavior observed in real-world neural networks.

- Biologically inspired: Attempting to mimic the behavior of biological neural systems in a way that is not necessarily faithful to biology.
- Neuron+Other: Neuron models have additional biologically inspired components, such as axons, dendrites, or glial cells, that are typically absent from other neuromorphic neuron models.
- Integrate-and-fire models are a subset of the more complex biologically inspired spiking neuron models.
- Models of neurons based on the McCulloch-Pitts neuron, which is widely utilized in the field of ANNs.

Equation for the output of Neuron x for the given model as:

$$M_x = f\left(\sum_{i=0}^{Y} wi, jni\right)$$

Mx is result of the equation, f is the function that is used for activation, Y is the count of input for neuron x, wi,j is the value from I to j, and ni is the result of neuron x.

Numerous hardware implementations of biologically credible and biologically motivated neuron models have been constructed. Cell membrane dynamics, which regulates factors like rate leakage throughout the neuron's biological membranes; ion stream dynamics, which regulates how ions move from and to a neuron, switching the neuron's charge level; axonal models, which may incorporate delay ingredients; and dendritic concepts, which regulate how many pre- and post-synapse neurons alter the current neuron. Izhikevich [24] provides a useful summary of various spiking neuron models.

1. Hodgkin-Huxley Model: Credible in terms of biology, the Hodgkin-Huxley model is widely accepted as the most credible biologically feasible model of a neuron. The Hodgkin-Huxley model, initially proposed in 1952, is a sophisticated neuron model that accounts for ion transport within and out of the neuron using four-dimensional nonlinear differential equations [46]. Due to their biological plausibility, Hodgkin-Huxley models have found widespread application in neuromorphic implementations that aim to faithfully simulate biological brain systems. The Morris Lecar model simplifies the model to a two-dimensional nonlinear equation, which is nonetheless physiologically reasonable. In neuroscience and neuromorphic systems, this model is widely used.

2. Biologically Inspired: The Fitzhugh-Nagumo and Hindmarsh-Rose simulations are two examples of reduced variations of the Hodgkin-Huxley system that have been executed in hardware. In an effort to represent behavior rather than aiming to imitate physical activity in biological systems, these models tend to be simpler computationally and simpler in terms of the number of features, while becoming more scientifically inspired than specific biological systems. Simplifying computation has the potential to yield more efficient and space-saving solutions in neuromorphic computing devices [47,48]. From the point of view of algorithms and learning methods, models with fewer parameters can be simpler to set and/or train.

3. Biologically Inspired Mechanisms (Neurons + Others): There are many other types of biologically inspired models out there. They typically include a lot more detail about living organisms than concepts found

in the machine learning and artificial intelligence literature, like membrane dynamics, modeling ion-channel dynamics, incorporating axons and/or dendrite models, and glial cell or astrocyte interactions.

Sometimes, brand-new models are made with the hardware in mind from the start. To facilitate implementation in low-power VLSI, on field-programmable gate array (FPGA), or using static complementary metal-oxide-semiconductor (CMOS), for example, equations based on the Fitzhugh-Nagumo, Morris Lecar, Hodgkin-Huxley, or other models have been modified or abstracted. The Hodgkin-Huxley model has also been revised by other researchers to incorporate more recent hardware advancements like the metal-oxide-semiconductor field-effect transistor (MOSFET) transistor and the single-electron transistor.

4. Integrate-and-Fire Neurons: The integrate-and-fire family of spiking neuron models ranges in complexity from the simplest (the basic integrate-and-fire) to that of the Izhikevich model and other complicated biologically inspired models. While less realistic from a biological perspective, integrate-and-fire models nonetheless provide sufficiently complicated behavior for application in spiking neural systems. Even in its simplest form, the integrate-and-fire model preserves the neuron's resting membrane potential. The probability on a neuron decays over duration in the faulty combine approach, which is an extension of the basic implementation. It's widely utilized in neuromorphic systems and considered a top model. The generic regressive combine method, along with the quadratic integrate-and-fire model utilized in some neuromorphic systems represents the next degree of complexity. The reactive exponential integrate-and-fire model adds another layer of complexity on top of the models already mentioned (e.g., the Izhikevich model). Similarly, neuromorphic systems have taken advantage of these.

There are currently digital adaptations of spiking neuron models, which complement the formerly analog-style spiking neuron models. Instead of using regressive or linear system of equations to describe the dynamics, a cellular automaton is typically used in digital spiking neuron models. For neuromorphic formulations, solutions of resonate-and-fire, and adaptations of rotate-and-fire digital spiking neurons, a hybrid analog/digital implementation was developed [49]. In order to display nonlinear response characteristics, a modified send information spiking model has been developed. Pulse-coupled networks have also made use of digital spiking neurons. The first hardware implementation of a neuron for a random neural network has been made.

Full network models will be referred to as "spiking neural networks" in the following sections. We make no claims about which specific neuron model is being used in the execution of these spiking networks. In addition, the neuron model is customizable in some hardware implementations like SpiNNaker, allowing for many neuron models to be realized within a single neuromorphic implementation.

5. When considering developing a reliable ANN, the standard McCulloch-Pitts neuron architecture is available in a wide range of forms. The processing element is a popular hardware version of the McCulloch-Pitts models. It uses a simple thresholding function as the activation function. Efforts have also been made to develop a system of different activation patterns for neurons based on the McCulloch-Pitts model. Although certain activation functions are computationally demanding, others have seen more success in the neural network domain. This computational burden can translate to configurability, and as a result, there are various activation functions and implementations that try to strike a balance between these two competing goals, as well as the model's overall accuracy and computing utility. Other hardware-based activation functions have included the ramp-saturation feature, linear, piecewise linear, step function, multi-threshold, radial basis function, the tangent sigmoid function, and periodic activation functions. The basic sigmoid function and the hyperbolic tangent function are the most certain circuits because they incorporate not only the sigmoid and hyperbolic tangent activation functions but also their derivatives because the gradient is used in the pull training process. Some implementations have centered on the generation of neurons with programmable activation functions or on the generation of building blocks for the generation of neurons.

Hardware implementations of neuron models have also been made for more classic forms of ANNs. There are various neuron models, such as those found in binary neural networks, fuzzy neural networks, and Hopfield neural networks. Generally speaking, many distinct neuron models have been realized in hardware, and one of the choices a user might make is a compromise between complexity and biological inspiration. There is a qualitative comparison between various neuron models with respect to these two criteria in Figure 10.4.

10.2.2 Synapse models

A significant amount of attention has been paid to generating synapse implementations separate from neuron models for neuromorphic systems, just as there has been a lot of attention paid to generating neuron models. Once again, we may break the synapse models into two groups: biologically inspired synapse implementations, which include synapses for spike-based systems, and synapse implementations for classic ANNs, including feedforward neural networks. It's important to remember that in neuromorphic systems, synapses are often the most numerous component, necessitating the greatest space on a given chip. For several practical systems and notably

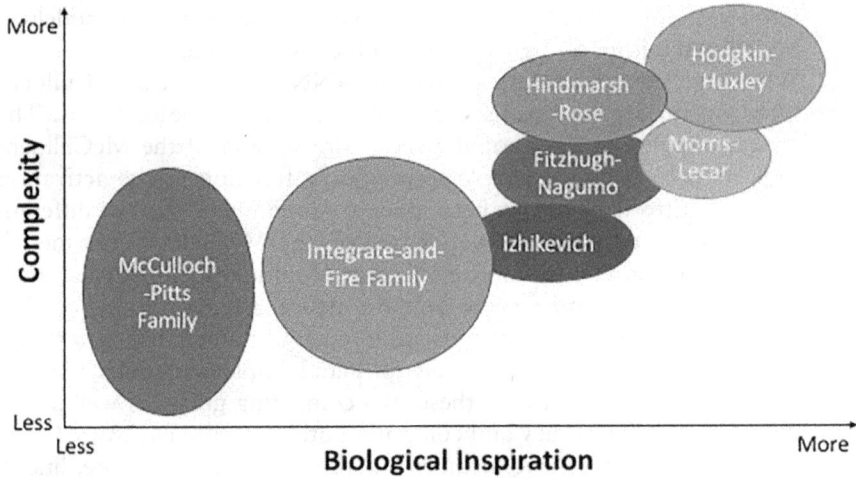

Figure 10.4 Graphical representation of neurons model comparison.

for the creation and usage of novel materials for neuromorphic, the focus is often on simplifying the synapse implementation. Therefore, unless an attempt is made to describe biological activity explicitly, synapse models tend to be quite straightforward. The strength or weight value of a neuron can alter over time thanks to a plasticity mechanism, which is often included in more complicated synapse models. In biological brains, mechanisms of plasticity have already been linked to knowledge.

10.2.2.1 Network model

Network models explain the relationships between neurons and synapses. The preceding paragraphs should have given you a sense of the breadth of the research into neural network models for computational models. Again, they span from statistically controlled, non-spiking neural networks to those that aim to emulate biological behavior as precisely as possible. When deciding on a network model, there are many aspects to take into account. As has been established in prior sections, one of the elements is undoubtedly the biological inspiration and complexity of neuron and synapse models. Network topology is another important aspect to think about. Network topologies that can be employed in different kinds of networks are illustrated in Figure 10.5. These topologies range from those influenced by biology to those based on electrical impulses. It's possible that the connectivity will be limited according on the devices selected, which will in turn limit the feasible topologies. Section will go into greater depth on a third consideration: whether or not the selected network model can make use of and benefit from preexisting training or learning methods. Finally, the

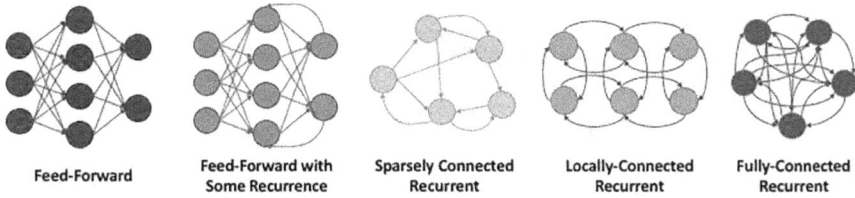

Figure 10.5 Network arrangement used for neuromorphic system.

network model's potential for broad use across a variety of applications is a factor that should not be discounted.

In hardware, conventional spiking neural networks come in a wide range of forms. The several types of integrate-and-fire neurons, as well as more physiologically plausible or biologically inspired models, are used in these implementations. The synapse architecture of a spiking biological system often includes spike-timing-dependent plasticity (STDP). Spiking models are often used in neuromorphic implementations due to their event-drive nature and superior energy efficiency compared to other systems. Thus, spiking neuromorphic systems have been used to implement various other types of neural network models, such as spiking feed-forward networks, spiking recurrent networks, spiking deep neural networks, spiking deep belief networks, spiking Hebbian systems, spiking Hopfield networks or associative memories. Another kind of neural network model was employed in these implementations, and it makes use of a spiking neural network architecture in neuromorphic devices. These techniques are often trained on a classic neural network architecture (e.g., feed-forward network) before the resulting network solution is modified to work with a spiking neuro-morphic implementation. As a result, the spiking neural network's full capabilities may not be exploited.

Figure 10.6 is a high-level depiction of the modeling approaches used in computational models. In Figure 10.7, we can see the development of several popular models used in neuromorphic implementations. According to the data, spiking and feed-forward implementations account for the vast majority of real-world examples, with spiking's popularity growing over the past decade. Convolutional neural networks in deep learning have become increasingly popular and fruitful over the past five years, while generalized feed-forward models have begun to decline.

Deep neural networks are one sort of artificial brain network that has frequently looked to sensory information for influence. Pulse-coupled human brains and cellular neural networks are two more models used in neuromorphic systems that were motivated by the visual system. While input signal networks were all the rage in the early 2000s, cellular neural networks were widely used in early neuromorphic implementations and are seeing a renaissance.

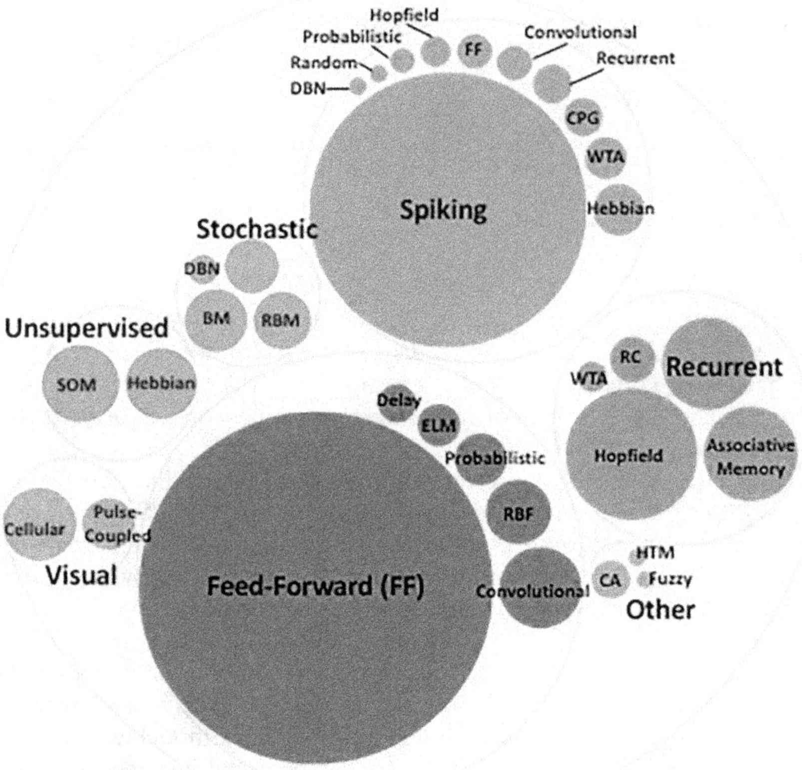

Figure 10.6 Basic neuromorphic application with network prototypical.

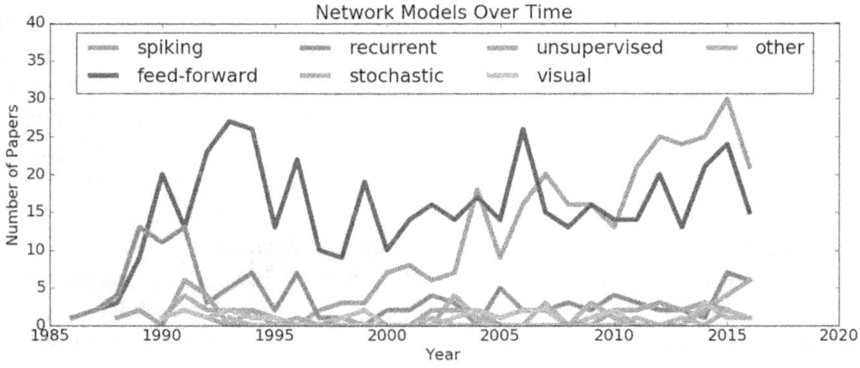

Figure 10.7 Neuromorphic execution over time with different models.

Cellular automata; fuzzy neural networks, which integrate fuzzy approach and ANNs; and the patriarchal temporal recollection schematic diagram introduced by Hawkins are all examples of less widely accepted neural network and neural network-neighboring models introduced in domain adaptation.

It is evident that there is a rich landscape of significant biological and artificial models for neural networks from which to choose when designing neuromorphic implementations. The goal of the neuromorphic system will significantly influence the selection of a suitable model. Models in neuroscience research projects tend to choose the side of being biologically plausible or at minimum biologically inspired. ANN-like technologies with shown strengths in these areas may be most relevant for algorithms that have been ported to equipment for a specific application, such as computer vision on a distant sensor or autonomous robots. For event-driven spiking neural network systems, it is common for the model to be selected or modified so that it fits within specific hardware features (e.g., picking models that employ STDP for memristors). The majority of neural network models have evidently been applied to hardware at a certain point in time.

10.3 ALGORITHM AND LEARNING

Procedures are at the heart of many of the outstanding concerns with neuromorphic systems. Selecting an appropriate algorithm is contingent on the neuron, synapse, and network models selected, while certain algorithms are tailored to particular network configurations, neuron simulations, or other aspects of modelling choices. In addition, a second concern is whether or not a system's retraining or learning should occur on-chip, or if systems should be learned off-chip and then transmitted to the neuromorphic architecture. Third, we must decide if online, unsupervised algorithms are required (in which case they must be implemented on-chip) or if off-line, supervised approaches are adequate, or if a hybrid approach is best. As a result of their potential for online learning, neuromorphic systems have gained popularity in the post-law Moore's era as a supplementary architecture. However, even the best-funded neuromorphic systems have had trouble developing algorithms for configuring their hardware. Algorithms that are developed specifically for hardware implementation, including those that run entirely on the chip or with the chip present, are the primary emphasis of this section, as shown in Figure 10.8.

10.3.1 Supervised learning concept

Back-propagation represents the most popular method for coding computational models. As a supervised learning technique, back-propagation isn't often considered an online approach. Feed-forward neural nets, recurrence

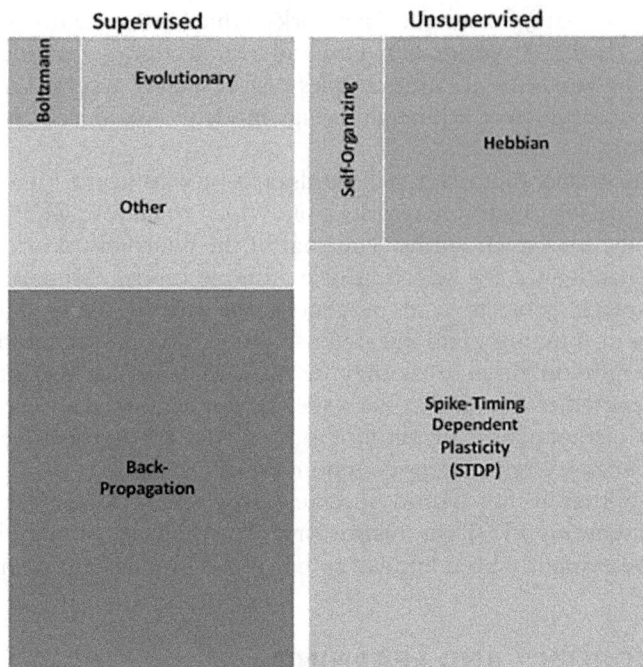

Figure 10.8 Step-by-step procedure of on-chip learning.

nets (often back-propagation through time), spiking nets (where often feed-forward nets are converted to spiking systems), and convolutional nets can all be trained with back-propagation and its many variants. Since there are numerous highly optimized software implementations available, using back-propagation off-line on a conventional host system is the simplest option. These methods mostly employ elementary back-propagation, which has been thoroughly explored elsewhere in the neural network literature.

10.4 CONCLUSION

The purpose of this paper is to provide a review of the literature on neuromorphic computing. Although the reasons for creating neuromorphic computers have evolved over time, the requirement for a non-von Neumann architecture that is low-power, massively parallel, able to perform in real time, and potentially capable of training or learning online has not. We went over several models of neurons, synapses, and networks that have been implemented in neuromorphic and neural network hardware before, stressing the wide range of options available when designing a neuromorphic system.

Each of these models has its own advantages and disadvantages, making it unlikely that they will ever be merged into a single, unified theory. It's safe to assume that anything from simple feed-forward neural networks to complex simulations of biological brain networks will continue to coexist in the neuromorphic computing ecosystem.

The many learning and training algorithms developed for and used by neuromorphic systems were discussed. Moving forward, we need to focus on developing dedicated training and learning algorithms for neuromorphic systems, as opposed to just adopting those built for other architectures. Our research suggests that this subfield of neuromorphic computing holds some of the greatest promise for future advancements. We talked about the big picture of neuromorphic system hardware and the cutting-edge device-level components and materials that are powering their development. As time goes on, there is also plenty of room for improvement here. We briefly covered some of the ancillary systems for neuromorphic computers, like ancillary software, of which there is relatively little and which would substantially assist the community. Finally, we went over some of the uses of neuromorphic computing systems.

With this work, we aimed to provide a comprehensive overview of neuromorphic computing research across multiple domains. Therefore, we have added all citations in this revision. The results of this study, we believe, will serve as motivation for others to create similarly novel approaches to help fill in the blanks with their own research and to think about how neuromorphic computers might work for their own purposes.

REFERENCES

[1] C. Mead, "Neuromorphic electronic systems," *Proceedings of the IEEE*, vol. 78, no. 10, pp. 1629–1636, Oct 1990.

[2] D. Monroe, "Neuromorphic computing gets ready for the (really) big time," *Communications of the ACM*, vol. 57, no. 6, pp. 13–15, 2014.

[3] J. Von Neumann and R. Kurzweil, *The computer and the brain*. Yale University Press, 2012.

[4] A. M. Turing, "Computing machinery and intelligence," *Mind*, vol. 59, no. 236, pp. 433–460, 1950.

[5] A. F. Murray and A. V. Smith, "Asynchronous vlsi neural networks using pulse-stream arithmetic," *Solid-State Circuits, IEEE Journal of*, vol. 23, no. 3, pp. 688–697, 1988.

[6] S. Lakkadi, A. Mishra, and M. Bhardwaj, "Security in ad hoc networks," *American Journal of Networks and Communications*, vol. 4, no. 3–1, pp. 27–34, 2015.

[7] I. Jain and Dr. M. Bhardwaj, "A Survey Analysis of COVID-19 Pandemic Using Machine Learning (July 14, 2022)," *Proceedings of the Advancement in Electronics & Communication Engineering*, 2022, Available at SSRN: https://ssrn.com/abstract=4159523 or http://dx.doi.org/10.2139/ssrn.4159523

[8] F. Blayo and P. Hurat, "A vlsi systolic array dedicated to hopfield neural network," in *VLSI for Artificial Intelligence*. Springer, 1989, pp. 255–264.

[9] F. Salam, "A model of neural circuits for programmable vlsi im- plementation," in *Circuits and Systems, 1989. IEEE International Symposium on*. IEEE, 1989, pp. 849–851.

[10] S. Bibyk, M. Ismail, T. Borgstrom, K. Adkins, R. Kaul, N. Khachab, and S. Dupuie, "Current-mode neural network building blocks for analog mos vlsi," in *Circuits and Systems, 1990. IEEE International Symposium on*. IEEE, 1990, pp. 3283–3285.

[11] A. Sharma, A. Tyagi, and M. Bhardwaj, "Analysis of techniques and attacking pattern in cyber security approach: A survey," *International Journal of Health Sciences*, vol. 6, no. S2, pp. 13779–13798, 2022. 10.53730/ ijhs.v6nS2.8625

[12] A. Tyagi, A. Sharma, and M. Bhardwaj, "Future of bioinformatics in India: A survey," *International Journal of Health Sciences*, vol. 6, no. S2, pp. 13767–13778, 2022. 10.53730/ijhs.v6nS2.8624.

[13] F. Distante, M. Sami, and G. S. Gajani, "A general configurable architecture for wsi implementation for neural nets," in *Wafer Scale Integration, 1990. Proceedings.,[2nd] International Conference on*. IEEE, 1990, pp. 116–123.

[14] J. B. Burr, "Digital neural network implementations," *Neural networks, Concepts, Applications, and Implementations*, vol. 3, pp. 237–285, 1991.

[15] M. Chiang, T. Lu, and J. Kuo, "Analogue adaptive neural network circuit," *IEE Proceedings G (Circuits, Devices and Systems)*, vol. 138, no. 6, pp. 717–723, 1991.

[16] K. Madani, P. Garda, E. Belhaire, and F. Devos, "Two analog counters for neural network implementation," *Solid-State Circuits, IEEE Journal of*, vol. 26, no. 7, pp. 966–974, 1991.

[17] A. F. Murray, D. Del Corso, and L. Tarassenko, "Pulse-stream vlsi neural networks mixing analog and digital techniques," *Neural Networks, IEEE Transactions on*, vol. 2, no. 2, pp. 193–204, 1991.

[18] P. Hasler and L. Akers, "Vlsi neural systems and circuits," in *Computers and Communications, 1990. Conference Proceedings, Ninth Annual International Phoenix Conference on*. IEEE, 1990, pp. 31–37.

[19] P. Chauhan, and M. Bhardwaj, "Analysis the Performance of Interconnection Network Topology C2 Torus Based on Two Dimensional Torus," *International Journal of Emerging Research in Management & Technology*, vol. 6, no. 6, pp. 169–173, 2017.

[20] N. S. Pourush, and M. Bhardwaj, "Enhanced Privacy-Preserving Multi-Keyword Ranked Search over Encrypted Cloud Data," *American Journal of Networks and Communications*, vol. 4, no. 3, pp. 25–31, 2015.

[21] J. Wu, S. A. Haider, M. Bhardwaj, A. Sharma, and P. Singhal, "Blockchain-Based Data Audit Mechanism for Integrity over Big Data Environments," *Security and Communication Networks*, vol. 2022, 2022.

[22] J.-C. Lee and B. J. Sheu, "Parallel digital image restoration using adaptive vlsi neural chips," in *Computer Design: VLSI in Computers and Processors, 1990. ICCD'90. Proceedings, 1990 IEEE International Conference on*. IEEE, 1990, pp. 126–129.

[23] L. Tarassenko, M. Brownlow, G. Marshall, J. Tombs, and A. Murray, "Real-time autonomous robot navigation using vlsi neural networks," in *Advances in neural information processing systems*, 1991, pp. 422–428.

[24] L. Akers, M. Walker, D. Ferry, and R. Grondin, "A limited- interconnect, highly layered synthetic neural architecture," in *VLSI for artificial intelligence*. Springer, 1989, pp. 218–226.

[25] P. H. Leong and M. A. Jabri, "A vlsi neural network for morphology classification," in *Neural Networks, 1992. IJCNN., International Joint Conference on*, vol. 2. IEEE, 1992, pp. 678–683.

[26] M. Bhardwaja and A. Ahlawat, "Evaluation of Maximum Lifetime Power Efficient Routing in Ad hoc Network Using Magnetic Resonance Concept," *Recent Patents on Engineering*, vol. 13, no. 3, pp. 256–260, 2019.

[27] M. Bhardwaj and A. Ahalawat, "Improvement of Lifespan of Ad hoc Network with Congestion Control and Magnetic Resonance Concept," in *International Conference on Innovative Computing and Communications*. Springer, Singapore, 2019, pp. 123–133.

[28] M. Bhardwaj and A. Ahlawat, "Optimization of Network Lifetime with Extreme Lifetime Control Proficient Steering Algorithm and Remote Power Transfer," *DEStech Transactions on Computer Science and Engineering*, 2017.

[29] G. Cairns and L. Tarassenko, "Learning with analogue vlsp mlps," in *Microelectronics for Neural Networks and Fuzzy Systems, 1994., Proceedings of the Fourth International Conference on*. IEEE, 1994, pp. 67–76.

[30] H. Markram, "The human brain project," *Scientific American*, vol. 306, no. 6, pp. 50–55, 2012.

[31] J. Schemmel, D. Bruderle, A. Grubl, M. Hock, K. Meier, and S. Millner, "A wafer-scale neuromorphic hardware system for large- scale neural modeling," in *Circuits and Systems (ISCAS), Proceedings of 2010 IEEE International Symposium on*. IEEE, 2010, pp. 1947–1950.

[32] J. Backus, "Can programming be liberated from the von neumann style?: a functional style and its algebra of programs," *Communications of the ACM*, vol. 21, no. 8, pp. 613–641, 1978.

[33] W. S. McCulloch and W. Pitts, "A logical calculus of the ideas immanent in nervous activity," *The Bulletin of Mathematical Biophysics*, vol. 5, no. 4, pp. 115–133, 1943.

[34] M. Bhardwaj and A. Ahlawat, "Enhance Lifespan of WSN Using Power Proficient Data Gathering Algorithm and WPT," *DEStech Transactions on Computer Science and Engineering*, 2017.

[35] M. Sharma, S. Rohilla, and M. Bhardwaj, "Efficient Routing with Reduced Routing Overhead and Retransmission of Manet," *American Journal of Networks and Communications. Special Issue: Ad Hoc Networks*, vol. 4, no. 3-1, pp. 22–26, May 2015. doi: 10.11648/j.ajnc.s.2015040301.15

[36] M. Bhardwaj, "7 Research on IoT Governance, Security, and Privacy Issues of Internet of Things," *Privacy Vulnerabilities and Data Security Challenges in the IoT*, p. 115, 2020.

[37] E. M. Izhikevich, "Which model to use for cortical spiking neurons?" *IEEE Transactions on Neural Networks*, vol. 15, no. 5, pp. 1063–1070, 2004.

[38] A. L. Hodgkin and A. F. Huxley, "A quantitative description of membrane current and its application to conduction and excitation in nerve," *The Journal of Physiology*, vol. 117, no. 4, p. 500, 1952.

[39] A. Basu, C. Petre, and P. Hasler, "Bifurcations in a silicon neuron," in *Circuits and Systems, 2008. ISCAS 2008. IEEE International Symposium on*. IEEE, 2008, pp. 428–431.

[40] F. Castanos and A. Franci, "The transition between tonic spiking and bursting in a six-transistor neuromorphic device," in *Electrical Engineering, Computing Science and Automatic Control (CCE), 2015 12th International Conference on*. IEEE, 2015, pp. 1–6.

[41] S. P. DeWeerth, M. S. Reid, E. A. Brown, and R. J. Butera Jr, "A comparative analysis of multi-conductance neuronal models in silico," *Biological Cybernetics*, vol. 96, no. 2, pp. 181–194, 2007.

[42] D. Dupeyron, S. Le Masson, Y. Deval, G. Le Masson, and J.-P. Dom, "A bicmos implementation of the hodgkin-huxley formalism," in *Microelectronics for Neural Networks, 1996., Proceedings of Fifth International Conference on*. IEEE, 1996, pp. 311–316.

[43] A. Kumar, S. Rohilla, and M. Bhardwaj, "Analysis of Cloud Computing Load Balancing Algorithms," *International Journal of Computer Sciences and Engineering*, vol. 7, pp. 359–362, 2019.

[44] M. Bhardwaj, A. Ahlawat, and N. Bansal, "Maximization of Lifetime of Wireless Sensor Network with Sensitive Power Dynamic Protocol," *International Journal of Engineering & Technology*, vol. 7, no. 3.12, pp. 380–383, 2018.

[45] M. Bhardwaj and A. Ahlawat, "Wireless Power Transmission with Short and Long Range Using Inductive Coil," *Wireless Engineering and Technology*, vol. 9, pp. 1–9, 2018. doi: 10.4236/wet.2018.91001.

[46] F. Grassia, T. Le'vi, S. Sa"ighi, and T. Kohno, "Bifurcation analysis in a silicon neuron," *Artificial Life and Robotics*, vol. 17, no. 1, pp. 53–58, 2012.

[47] M. Grattarola, M. Bove, S. Martinoia, and G. Massobrio, "Silicon neuron simulation with spice: Tool for neurobiology and neural networks," *Medical and Biological Engineering and Computing*, vol. 33, no. 4, pp. 533–536, 1995.

[48] A. M. Hegab, N. M. Salem, A. G. Radwan, and L. Chua, "Neuron model with simplified memristive ionic channels," *International Journal of Bifurcation and Chaos*, vol. 25, no. 06, p. 1530017, 2015.

[49] K. M. Hynna and K. Boahen, "Thermodynamically equivalent silicon models of voltage-dependent ion channels," *Neural Computation*, vol. 19, no. 2, pp. 327–350, 2007.

Chapter 11

Analysis of technology research and ADHD with the neurodivergent reader

A survey

Manish Bhardwaj[1], Jyoti Sharma[2], Analp Pathak[2], Vinay Kumar Sharma[3], and Mayank Tyagi[2]

[1]Department of Computer Science and Information Technology, KIET Group of Institutions, Delhi-NCR, Ghaziabad, India

[2]Department of Information Technology, KIET Group of Institutions, Delhi-NCR, Ghaziabad, India

[3]School of Computing Science and Engineering, Galgotias University, Greater Noida, India

11.1 INTRODUCTION

The word "neurodiversity" describes the wide variation in people's neurological make-ups, with a focus on the many different ways people think. This movement is driven by self-advocates and opposes the deficit or impairment labeling of neurotypes [1].

This was originally brought to human–computer interaction (HCI) by Dalton, who, in discussing the possibilities of the notion of neurodiversity for technology design and research, urged HCI researchers to work with neurodiverse populations and to recognize and promote the talents of individuals with such differences [2–4]. Neurodiversity, as used here, is the lived experience of cognitive and/or expressive differences from the neurotypical (including but not limited to medical designations such as autism, dyslexia, and attention deficit hyperactivity disorder [ADHD]).

Similarly, Mankoff et al. advocated for research on assistive technology to be guided by the lens of disability studies and to actively connect with disability communities. So, there has been a rise in research on neurodivergent populations, and as a result, neurodiverse academics have contributed to a better understanding of the language used in the fields of computer science and HCI [5]. First recommendations for the design of technology related to ADHD already exist, but primarily from a medically educated perspective.

Recent work by Cibrian et al. provides an overview of tools that can help children with ADHD learn to self-regulate [6,7]. What has been lacking up to this point is a critical investigation that focuses on how technological

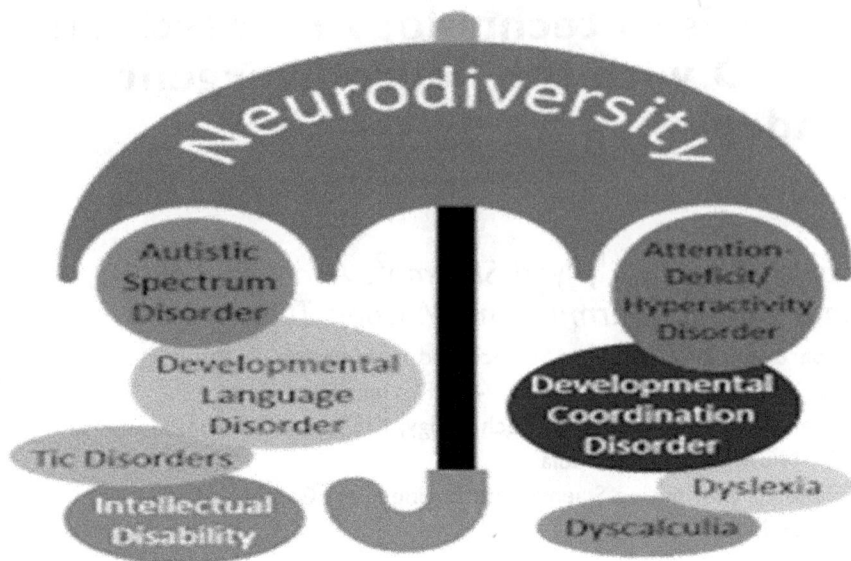

Figure 11.1 Basic conditions under neurodiversity.

research occurs in the context of ADHD, especially from a perspective expressly molded by persons with ADHD themselves.

The first-person perspective of people with disabilities (whether researchers or not) is crucial to understanding how these technologies work, and while modern HCI research in the realm of technologies and disabilities is increasingly informed by theories and practices from disability studies, this is not always the case [8,9]. Although autoethnographic studies and research focused on agency and self-determination have recently entered the field, these viewpoints have historically been less prevalent. Figure 11.1 shows the conditions that are present under one name "neurodiversity".

We contribute to this expanding body of work by conducting a systematic literature review of HCI and computer science studies that focus on ADHD [10]. This is consistent with earlier studies of technology and neurodivergence or disabilities more generally, such as a review by Spiel et al. of technologies for autistic children, a survey by Mack et al. of accessibility research, studies of wearables for autism intervention, and studies of research pertaining to neurodivergence and play [11].

Moreover, our work is a reaction to and an expansion on prior recommendations for learning prospective design techniques relating to technologies for people with ADHD [12]. Since neurodivergent people are often left out of studies focusing on them, we think it's important to expressly adopt this subjective stance [13]. This level of participation, however, affects how we read, analyze, and understand the corpus content. So, our emotional

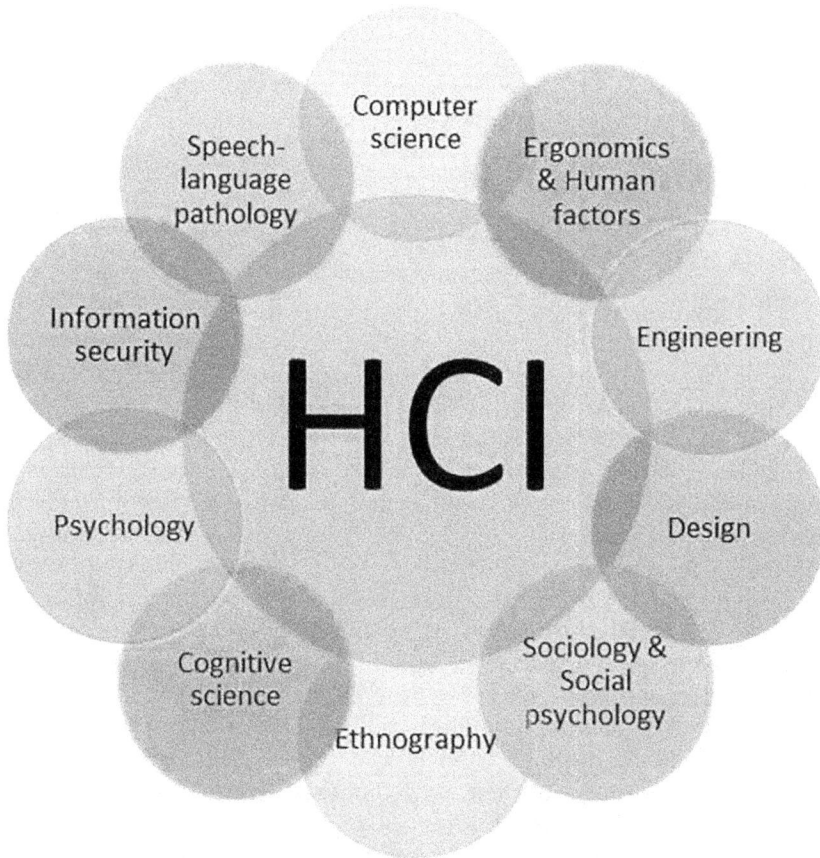

Figure 11.2 Research fields related to human–computer interaction.

reactions to how our coworkers characterize us, our communities, and our loved ones play a significant role in our writing [14].

We now provide context for our study by outlining our knowledge of ADHD, describing our previous work in the context of HCI and neurodiversity, and introducing our theoretical support from Critical Disability Studies and, more specifically, Crip Technoscience. Figure 11.2 shows the various research fields that are related to the HCI field.

After that, we explain our methodology in further detail. Our analysis and results show how existing research exposes ADHD as a problem space for technology design due to solutionist and paternalistic perspectives of the intended audience [15]. We draw conclusions for the technological research community, propose hypotheses about possible solutions, consider the consequences of engaging with works on a deeply personal level, and sketch the contours of possible future developments based on these results [16–18].

11.2 DOMAIN PREVIOUS STUDY

First, we explain ADHD briefly while addressing the ongoing research in the field of HCI and neurodivergence [19], before our discussion of relevant hypotheses building on Radical Theory and practice and specifically the concept of "Crip Technoscience".

11.2.1 ADHD

Clinically, hyperactivity, impulsivity, and inattention are the hallmarks of ADHD. Predominant inattentiveness [20–22], hyperactive-impulsiveness, and a mixed profile are frequently used diagnostic criteria [ibid]. For a long time, ADHD, or hyperkinetic behavior syndrome, was thought of as a condition that only affected kids [23].

Because of its early definition as a childhood condition, adults who sought a diagnosis were overlooked, and its failure to account for the "inattentive" (daydreamer) type led to a long-held misconception that ADHD mostly affects boys [24].

Research suggests that the prevalence of ADHD in the general population diminishes progressively throughout age groups, lending credence to the idea that one might "grow out of" the disorder [25].

This misconception persists despite growing evidence that the under-lying neurological distinctions that define ADHD persist over the course of a person's lifetime [26–29]. ADHD may be less noticeable in adults because of differences in their environments (work, family life, etc.; whereas in most countries, children all go to school), and because they have learned to adapt to their surroundings through time, using a variety of contextual masking tactics.

Misdiagnosis of depression or oppositional defiant disorder might occur because of the presentation of traits being connected to conventional standards along gender and race.

In recent years, those who have been diagnosed with ADHD have contributed to our growing body of knowledge about what it's like to live with and manage ADHD in the real world.

Research that takes both a critical and appreciative stance and the participation of people with ADHD in study both add to the credibility of these narratives [30–33]. In this context, lobbying focuses mostly on dispelling harmful myths and rebutting internalized societal stigma.

11.2.2 Neurodivergence and HCI

Proponents in HCI often claim to do accessibility research, although the precise nature of this work is rarely defined or agreed upon. Mack et al. found that only around 5% of self-identified accessibility research publications published between 1994 and 2019 focused on neurodivergent communities (which includes autistic populations).

In a similar vein, a recent analysis by Spiel and Gerling on HCI games research and neurodivergent communities found that a disproportionate amount of attention was paid to autistic populations compared to those with other conditions [34].

As a result, autism spectrum disorder (ASD) and children with ASD are common topics of discussion in the existing reviews in this field. One review, driven primarily by a medical and deficiency-oriented approach [35], provides implications for technological research in the setting of ADHD, and the other, focused on children with ADHD, does the same.

Here, we offer a quick recap of the current literature devoted to designing technology for people with neurodiversities. This study includes instances in which neurodivergent people were involved in the design of data visualizations due to nonstandard processing of visual information or the creation of shared spaces in which neurotypical and neurodivergent people can meet [36], for example, in the context of children's play or adults in the context of the workplace. Figure 11.3 shows the graphical representation of neurodiversity between low cognative and high cognative ability.

Self-determined design experiments are rarely seen in published work, but one such example is Damiani's creative examination of the particulars of neurodivergent embodiments.

We see these as crucial examples of what research and design may look like when driven by neurodiverse people, and this bodes well for future

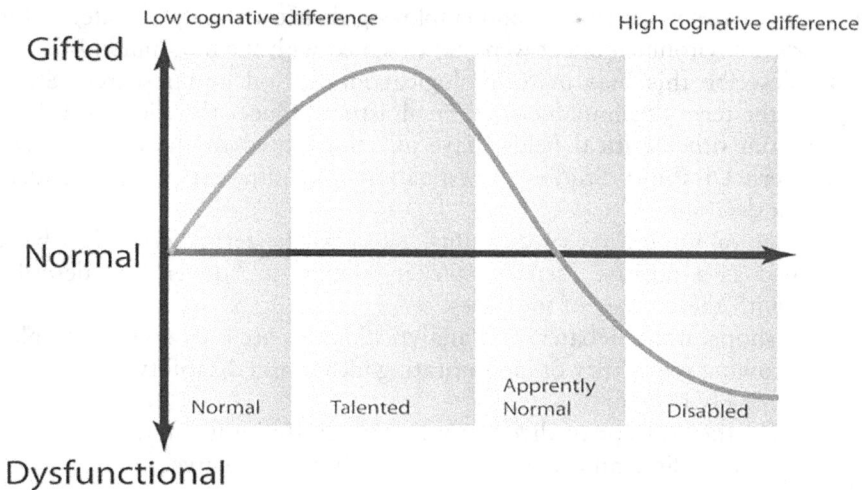

Figure 11.3 Graphical representation of neurodiversity between low cognative and high cognative ability.

studies on ADHD. Following the proverb in disability activism that claims "Nothing about us without us," Parsons et al. show the knowledge- and human-rights-based implications of including autistic people in research about them [37,38], which is a response to the overwhelming amount of research being driven from the outside in the context of autism.

They provide a real example to show how one author's declaration of an ADHD diagnosis led to them being disqualified from consideration for an otherwise suitable job posting [ibid].

For both neurotypical and neurodiverse academics, the inclusion of neurodiverse communities in their studies has important consequences for the negotiation of a shared impaired (or nondisabled) identity.

However, little attention has been paid to the potential forms that self-determined and participatory research can take in the context of ADHD [39], and a few of researchers who have ADHD have chosen not to declare their condition publicly.

As a result, our study contributes to the literature on neurodiversity and technology development by offering a dedicated analysis to the ADHD setting carried out by writers who do so expressly as impacted parties.

11.2.3 Studies of disabilities with a critical lens and crip technology

Disability Studies was first introduced to HCI scholars by Mankoff et al. in 2010. To be more specific, the autonomous premise of Disability Studies provides a critical counterpoint to the common understanding of assistive technologies for people with disabilities, which sees them as devices whose primary purpose is to maintain a corporeal standard, rather than to aid people with disabilities in developing and employing their own unique strategies for navigating environments that were not designed with them in mind [40].

To describe this bias in technological innovation and research, Shew created the term "technoableism". Feminist and Queer theories, as well as those from other critical fields, have all contributed to the development of modern Disability Studies' investigation into different ways of understanding disability.

Because of the efforts of disability activists, the term "crip" has been reclaimed as a positive identity marker (similar to "queer") to describe people with these types of identities.

Workshops, lively debates, and analytical lenses are only some examples of the growing popularity of incorporating ideas from disability studies into the field of HCI.

As such, the concept of diverse models of disability is important to our work, positionality, and analysis. Many other perspectives exist on disability, but the most common divide is between the medical (or deficiency-oriented) model, which emphasizes the individual, and the social (or access-oriented) model, which emphasizes structural impediments.

Living with an ADHD neurotype, however, is characterized by a discrepancy between internal expectations and the way in which the individual processes information from the outside world.

Disabling experiences become ingrained in our life alongside the widespread stigma of declaring ADHD and requesting adjustments. So, we have chosen to use the term "disability" in order to show our support for those who share these experiences, both as a person looking for crip kinship and as a person describing aspects of our own identity.

Because not doing so or working from a medical viewpoint on disabilities is itself a political act – any presumption of a default status is fundamentally political – we interact with this identity as disabled on purpose, despite the fact that doing so is a political act.

Hamraie and Fritsch provide a variety of perspectives on the nature of these technologies within the framework of viewing disability, and by extension, the space in which people with disabilities live, as political. Their Crip Technoscience framework prioritizes people with disabilities by pledging to increase accessibility, encourage collaboration, and promote disability justice. While HCI has been exposed to the concept of neurodiversity, no prior analysis of Crip Technoscience has focused on its relevance to neurodivergent populations.

11.3 RESEARCH METHODS

This literature evaluation took around 2.5 years to complete, beginning in early 2019.

We show how we got here by first describing the reading perspective we used, which was also the driving force behind our study.

Furthermore, we detail the methods we used to build and analyze the corpus as a whole. All of the writers' professional interest in technology research pertaining to ADHD inspired this undertaking. The first author reached out to other neurodiverse people to talk about problems they had noticed in self-descriptions.

We all got together at a conference at the beginning of 2019 to introduce ourselves to one another and launch the project. We also passed out reading glasses and established preliminary areas of interest at this time.

Our goal in demonstrating the relevance of this work to people with ADHD is, in part, to "support the construction of 'safer' settings for marginalized academics and students," in this case, neurodiverse peers.

Following Kafer's lead in adopting crip time, which "bends the clock to meet crippled bodies and minds" instead of vice versa, we spread this effort out over the course of 2.5 years to make explicit space for multiple temporalities.

Moreover, we must recognize that we had to factor in sorrow time into this crip period, as we lost peers and friends or had to deal with the repercussions of a global pandemic, including individual disease, ourselves.

11.3.1 Development of corpus

We built not one, but two corpora, the main one and an expansion on top of it. Our combined corpus excludes workshop descriptions, panel proposals, and journal articles, but includes full/long papers, short papers, works-in-progress and posters, demonstrations, and journal articles reporting on completed research.

On February 19, 2019, we compiled the initial core corpus by searching for the terms "ADHD" and "ADD" in the titles, abstracts, and keywords of publications in the ACM *Digital Library Guide to Computing Literature* and the hcibib indexing platform (although the latter stopped indexing content in 2018).

For this reason, we have chosen to restrict our review to papers dealing with computing and HCI. As a result, 56 papers were produced. After reading each abstract, we culled the corpus down to 52 papers that directly addressed our research questions. We refrained from making any more cuts to the corpus, which enabled us to compare and contrast longer and shorter publications and to report on current trends. The use of merely these two references suggests that we have not made a concerted effort to read through the technical articles published in other clinical or medical journals. Yet, we zero in on the ways in which computing, and especially HCI-oriented literature, constructs ADHD and related technologies.

Afterwards, we surveyed more recent papers from the same source, covering the time period from around the middle of 2018 to the end of 2020, to see if the patterns we had noticed in the original corpus were still there, and/or if any new patterns emerged.

One workshop, one paper written in a language none of the authors can read fluently, and full online versions of PhD theses were all dismissed as irrelevant to our investigation. This process added 48 more publications to the 'expanded corpus' (after duplicates were removed). With the growing body of literature on neurodiversity in HCI, we decided to do a broad survey of these additional studies to test the validity of our earlier conclusions.

11.4 RESULT AND DISCUSSION

Each of the four authors read the core corpus with a focus on either participants (i.e., who was included or addressed in the research and how did potential participants behave), disability (i.e., how did authors conceptualize and explain ADHD), researchers (i.e., the larger research framing and disciplinary origin of the work), or technology (i.e., the design and development processes, including the artifacts and their purpose).

To a) facilitate chunky reading in a systematic way and b) permit possibly divergent reading of the same publications to correctly account for varied strengths in the research we surveyed, we defined separate reading lenses.

We then compared and contextualized our studies across different authors by using a procedure of procedural coding along subcategories of these focuses. The authors addressed several papers that had received contrasting ratings after perusing each other's remarks and codes. The goal here was not to force consistency among the many evaluations, but rather to make effective use of them in order to better comprehend the paper's merits and shortcomings.

In order to conduct our theoretical investigation, we first created a conceptual map of the publications in the corpus. After that, we used Critical Disability Studies theories to analyze the publications' implications from the neurodiverse perspective. We followed Boyatzis' method of theme analysis, which permits inductive and deductive coding by numerous coders while taking into account the situatedness and subjective quality standards of acceptable categorization.

We avoided a codebook method, which is more numerically focused, on purpose so that we could include more voices in our study and bolster our findings by working through our disagreements.

Two of the four original coders similarly analyzed the expanded corpus to spot themes in recent research that compared to our initial findings. With the exception of when a new pattern emerged, this study relied on previously defined codes. Discordant findings are presented in separate paragraphs adjacent to the relevant sections. Many of the papers in this extended corpus are marked as part of the "extended corpus" because they are excellent illustrations of patterns observed in the main corpus.

We make no assumptions about the neurotypes of the people who participated in the research that is reflected in our database. Indeed, we have a keen awareness of the potential adverse effects of disclosure. Instead, we evaluate the research strategy and viewpoint taken from a discursive perspective.

To paraphrase Haraway, we intend to "take the privilege of our limited perspective[s]" seriously, rather than make generalizations based on our reading.

Our findings highlight the prevalence of tools for intervention and diagnosis in the field of ADHD. We also found that those with ADHD are often left out of the loop when it comes to the development of tools meant specifically for them. This results in a solitary framing of the target population as a source of "issues to address" and technology that predominantly mirror neurotypical expectations rather than neurodivergent wants and desires.

The implications of these findings for identifying the "user" of a given technology in neurodiverse settings are discussed, along with what Crip Technoscience-based alternatives to current technologies would entail.

11.5 RESEARCH GAPS

In addition to the obvious implications for the field of ADHD technology research as a whole, we also found a number of open areas where additional investigation into the role(s) technologies play in this setting would be beneficial.

Our presentation of these alternatives is meant to be illustrative and exploratory, and not exhaustive. It does reveal, however, that there are fundamental gaps in the knowledge production about ADHD and technologies, particularly in regards to language, the inclusion of persons with ADHD, and a closer examination of the discursive and material repercussions that technology embodies.

Most of the studies we looked at were conducted on children, and even those that weren't frequently employed children or adolescents as subjects (i.e., children or adolescents).

The lack of research involving adults, and especially older persons with ADHD, stands out to us as a major hole. The function that technology plays in an adult's life is likely to change from that of a child's since adults have less external structure in their lives. Also, compared to adults, children are given less credit for having the ability to make their own choices regarding their interactions with technology and are more likely to become embedded in existing power systems.

As a result, purposeful interaction with adults has the potential to undermine the paternalistic assumptions that now underpin much technological investigation into ADHD.

We also see promise in tools that enhance executive function during freely chosen activities. Instead of forcing persons with ADHD to follow a predetermined set of tasks, we propose looking for ways to help them figure out how to accomplish the goals they have for themselves.

Finally, we think there's value in studying how persons with ADHD are adapting and using current technologies. One way this could manifest itself is in the way these people use social media as a means to both find and share information and form groups centered on the construction of personal meaning.

There is also a lack of research on why many life-organization methods that are effective for neurotypical people are ineffective for people with ADHD (e.g., weekly calendar planning).

Since we recognize that researchers in this field of study must first establish a foundation of trust, we call on our peers and colleagues alike to help bring this potential to fruition.

Here, we're hoping that our work might help lay the groundwork for a fruitful discussion.

11.6 CONCLUSION

From the perspective of neurodiverse readers, we analyzed works concerning technology in the context of ADHD. Our findings highlight the barriers that prevent people with ADHD from participating in the design of aiding technologies.

We also found that until recently, researchers' perspectives remained static, with the exception of a shift toward a concentration on diagnostic and interventionist approaches in technology development.

Consequences of the current research's reliance on a deficit model of ADHD features were then discussed. We next provide a number of suggestions for developments in this area based on our speculative choices. In this way, we advocate for researchers to abandon their adherence to neurotypical standards and approach their work with us and other marginalized communities less from a paternalistic and more from a solidaric and community-oriented stance.

There are constraints on this work, like there are on any. For one, all of us writing this are white, we live and work in the Global North, and, as we've already mentioned, we have the advantage of having access to a diagnosis, which in turn governs our ability to receive social accommodations.

As a matter of fact, we owe our academic success to the fortunate circumstances of our own educational backgrounds. We can thus only speak for ourselves in our mutual assessment and make no claims to represent all people with ADHD. However, we provide a close reading of existing works and outline the future potential of such appreciative approaches in light of the growing number of researchers who openly disclose their ADHD, also within HCI, and the emerging research into self-determined options for technologies in this context.

We can't exist without technology, but if it's only used to sort individuals into boxes and control their actions, we have a responsibility as technologists to find better uses for it.

REFERENCES

[1] Ashwag Al-Shathri, Areej Al-Wabil, and Yousef Al-Ohali. 2013. Eye-Controlled Games for Behavioral Therapy of Attention Deficit Disorders. In *HCI International 2013 – Posters' Extended Abstracts*, Constantine Stephanidis (Ed.). Springer Berlin Heidelberg, Berlin, Heidelberg, 574–578.

[2] Shahab U. Ansari. 2010. *Validation of FS+LDDMM by Automatic Segmentation of Caudate Nucleus in Brain MRI*. In Proceedings of the 8th International Conference on Frontiers of Information Technology (Islamabad, Pakistan) (FIT '10). Association for Computing Machinery, New York, NY, USA, Article 10, 6 pages. 10.1145/1943628.1943638

[3] J. Anuradha, Tisha, Varun Ramachandran, K. V. Arulalan, and B. K. Tripathy. 2010. Diagnosis of ADHD Using SVM Algorithm. In Proceedings of the Third Annual ACM Bangalore Conference (Bangalore, India) (COMPUTE '10). Association for Computing Machinery, New York, NY, USA, Article 29, 4 pages. 10.1145/1754288.1754317.

[4] A. Kumar, S. Rohilla, and M. Bhardwaj (2019). Analysis of Cloud Computing Load Balancing Algorithms. *International Journal of Computer Sciences and Engineering*, 7, 359–362.

[5] M. Bhardwaj, A. Ahlawat, and N. Bansal (2018). Maximization of Lifetime of Wireless Sensor Network with Sensitive Power Dynamic Protocol. *International Journal of Engineering & Technology*, 7(3.12), 380–383.

[6] M. Bhardwaj and A. Ahlawat (2018) Wireless Power Transmission with Short and Long Range Using Inductive Coil. *Wireless Engineering and Technology*, 9, 1–9. doi: 10.4236/wet.2018.91001.

[7] Othman Asiry, Haifeng Shen, and Paul Calder. 2015. Extending Attention Span of ADHD Children through an Eye Tracker Directed Adaptive User Interface. In Proceedings of the ASWEC 2015 24th Australasian Software Engineering Conference (ASWEC' 15 Vol. II). Association for Computing Machinery, New York, NY, USA, 149–152. 10.1145/2811681.2824997

[8] Kathleen B Aspiranti and David M Hulac. 2021. Using Fidget Spinners to Improve On-Task Classroom Behavior for Students With ADHD. *Behavior Analysis in Practice* (2021), 1–12.

[9] Peter Balan, Eva Balan-Vnuk, Mike Metcalfe, and Noel Lindsay. 2015. Concept mapping as a methodical and transparent data analysis process. In *Handbook of Qualitative Organizational Research*. Routledge, 350–362.

[10] Kess L. Ballentine. 2019. Understanding Racial Differences in Diagnosing ODD Versus ADHD Using Critical Race Theory. *Families in Society* 100, 3 (2019), 282–292. 10.1177/1044389419842765 arXiv:https://doi.org/10.1177/1044389419842765

[11] M. Bhardwaj and A. Ahlawat (2017). Enhance Lifespan of WSN Using Power Proficient Data Gathering Algorithm and WPT. *DEStech Transactions on Computer Science and Engineering*.

[12] Megha Sharma, Shivani Rohilla, and Manish Bhardwaj, Efficient Routing with Reduced Routing Overhead and Retransmission of Manet, *American Journal of Networks and Communications*. Special Issue: Ad Hoc Networks. Volume 4, Issue 3–1, May 2015, pp. 22–26. doi: 10.11648/j.ajnc.s.2015040301.15

[13] M. Bhardwaj (2020). 7 Research on IoT Governance, Security, and Privacy Issues of Internet of Things. *Privacy Vulnerabilities and Data Security Challenges in the IoT*, 115.

[14] Elizabeth Barnes. 2016. *The minority body: A theory of disability*. Oxford University Press.

[15] M. Á. Bautista, A. Hernández-Vela, S. Escalera, L. Igual, O. Pujol, J. Moya, V. Violant, and M. T. Anguera. 2016. A Gesture Recognition System for Detecting Behavioral Patterns of ADHD. *IEEE Transactions on Cybernetics* 46, 1 (2016), 136–147.

[16] Robert Beaton, Ryan Merkel, Jayanth Prathipati, Andrew Weckstein, and Scott McCrickard. 2014. Tracking Mental Engagement: A Tool for Young People with ADD and ADHD. In Proceedings of the 16th International ACM SIGACCESS Conference on Computers & Accessibility (Rochester, New York, USA) (ASSETS '14). Association for Computing Machinery, New York, NY, USA, 279–280. 10.1145/2661334.2661399

[17] M. Bhardwaja and A. Ahlawat (2019). Evaluation of Maximum Lifetime Power Efficient Routing in Ad hoc Network Using Magnetic Resonance Concept. *Recent Patents on Engineering*, 13(3), 256–260.

[18] M. Bhardwaj and A. Ahalawat (2019). Improvement of Lifespan of Ad hoc Network with Congestion Control and Magnetic Resonance Concept. In *International Conference on Innovative Computing and Communications* (pp. 123–133). Springer, Singapore.

[19] M. Bhardwaj and A. Ahlawat (2017). Optimization of Network Lifetime with Extreme Lifetime Control Proficient Steering Algorithm and Remote Power Transfer. *DEStech Transactions on Computer Science and Engineering*.

[20] P. Chauhan and M. Bhardwaj (2017). Analysis the Performance of Interconnection Network Topology C2 Torus Based on Two Dimensional Torus. *International Journal of Emerging Research in Management & Technology*, 6(6), 169–173.

[21] Cynthia L. Bennett, Erin Brady, and Stacy M. Branham. 2018. Inter-dependence as a Frame for Assistive Technology Research and Design. In Proceedings of the 20th International ACM SIGACCESS Conference on Computers and Accessibility (Galway, Ireland) (ASSETS '18). Association for Computing Machinery, New York, NY, USA, 161–173. 10.1145/3234695.3236348

[22] Cynthia L. Bennett and Daniela K. Rosner. 2019. *The Promise of Empathy: Design, Disability, and Knowing the "Other"*. Association for Computing Machinery, New York, NY, USA, 1–13. 10.1145/3290605.3300528

[23] Jared David Berezin. 2014. Disabled capital: A narrative of attention deficit disorder in the classroom through the lens of Bourdieu's capital. *Disability Studies Quarterly* 34, 4 (2014).

[24] K Birth. 2017. *Time Blind*. Springer.

[25] N. S. Pourush and M. Bhardwaj (2015). Enhanced Privacy-Preserving Multi-Keyword Ranked Search over Encrypted Cloud Data. *American Journal of Networks and Communications*, 4(3), 25–31.

[26] J. Wu, S. A. Haider, M. Bhardwaj, A. Sharma, and P. Singhal (2022). Blockchain-Based Data Audit Mechanism for Integrity over Big Data Environments. *Security and Communication Networks*, 2022.

[27] A. Sharma, A. Tyagi, and M. Bhardwaj (2022). Analysis of techniques and attacking pattern in cyber security approach: A survey. *International Journal of Health Sciences*, 6(S2), 13779–13798. 10.53730/ijhs.v6nS2.8625

[28] Mark Blythe, Jamie Steane, Jenny Roe, and Caroline Oliver. 2015. *Solutionism, the Game: Design Fictions for Positive Aging*. Association for Computing Machinery, New York, NY, USA, 3849–3858. 10.1145/2702123.2702491

[29] Natale Salvatore Bonfiglio, Roberta Renati, Davide Parodi, Eliano Pessa, Dolores Rollo, and Maria Pietronilla Penna. 2020. Use of training with BCI (Brain Computer Interface) in the management of impulsivity. In 2020 IEEE international symposium on medical measurements and applications (MeMeA). IEEE, 1–5.

[30] LouAnne Boyd. 2019. Designing Sensory-Inclusive Virtual Play Spaces for Children. In Proceedings of the 18th ACM International Conference on Interaction Design and Children. 446–451.

[31] LouAnne Boyd, Kendra Day, Ben Wasserman, Kaitlyn Abdo, Gillian Hayes, and Erik Linstead. 2019. Paper Prototyping Comfortable VR Play for Diverse Sensory Needs. In Extended Abstracts of the 2019 CHI Conference on Human Factors in Computing Systems (CHI EA '19). Association for Computing Machinery, New York, NY, USA, 1–6. 10.1145/3290607.3313080

[32] Emeline Brulé and Katta Spiel. 2019. Negotiating Gender and Disability Identities in Participatory Design. In Proceedings of the 9th International Conference on Communities & Technologies - Transforming Communities (Vienna, Austria) (C&T '19). Association for Computing Machinery, New York, NY, USA, 218–227. 10.1145/3328320.3328369

[33] Fiona Campbell. 2009. *Contours of ableism: The production of disability and abledness*. Springer.

[34] A. Tyagi, A. Sharma, and M. Bhardwaj (2022). Future of bioinformatics in India: A survey. *International Journal of Health Sciences*, 6(S2), 13767–13778. 10.53730/ijhs.v6nS2.8624.

[35] S. Lakkadi, A. Mishra, and M. Bhardwaj (2015). Security in ad hoc networks. *American Journal of Networks and Communications*, 4(3-1), 27–34.

[36] Jain, Ishita and Dr. Manish Bhardwaj, A Survey Analysis of COVID-19 Pandemic Using Machine Learning (July 14, 2022). Proceedings of the Advancement in Electronics & Communication Engineering 2022, Available at SSRN: https://ssrn.com/abstract=4159523 or http://dx.doi.org/10.2139/ssrn.4159523

[37] Will H Canu, Matthew L Newman, Tara L Morrow, and Daniel L. W Pope. 2008. Social Appraisal of Adult ADHD: Stigma and Influences of the Beholder's Big Five Personality Traits. *Journal of Attention Disorders* 11, 6 (2008), 700–710.

[38] Dario Cazzato, Silvia M Castro, Osvaldo Agamennoni, Gerardo Fernández, and Holger Voos. 2019. A Non-Invasive Tool for Attention-Deficit Disorder Analysis Based on Gaze Tracks. In Proceedings of the 2nd International Conference on Applications of Intelligent Systems (Las Palmas de Gran Canaria, Spain) (APPIS '19). Association for Computing Machinery, New York, NY, USA, Article 5, 6 pages. 10.1145/3309772.3309777

[39] Kuo-Chung Chu, Hsin-Jou Huang, and Yu-Shu Huang. 2016. Machine learning approach for distinction of ADHD and OSA. In 2016 IEEE/ACM international conference on advances in social networks analysis and mining (ASONAM). IEEE, 1044–1049.

[40] Franceli L Cibrian, Kimberley D Lakes, Sabrina Schuck, Arya Tavakoulnia, Kayla Guzman, and Gillian Hayes. 2019. Balancing caregivers and children interaction to support the development of self-regulation skills using a smartwatch application. In Adjunct Proceedings of the 2019 ACM International Joint Conference on Pervasive and Ubiquitous Computing and Proceedings of the 2019 ACM International Symposium on Wearable Computers. 459–460.

Index

For Product Safety Concerns and Information please contact our EU
representative GPSR@taylorandfrancis.com
Taylor & Francis Verlag GmbH, Kaufingerstraße 24, 80331 München, Germany

www.ingramcontent.com/pod-product-compliance
Lightning Source LLC
Chambersburg PA
CBHW060405220326
41598CB00023B/3019